The Earth's Killer C's

AN ECO-CONSERVATIVES EASY GUIDE TO THE ENVIRONMENTAL CRISIS

Peter Gwillim Kreitler

Cover
100% Post Consumer Waste Non Varnished
Recycled Paper
Text
100% kenaf plant paper
Ink
Soy Based

ISBN 0 - 9629069-5-6

Peter G. Kreitler, Morning Sun Press,
and Alonzo Printing have sought to produce
a product that is ecologically sound so as to
minimize the impact on our environment.

Individual copies: $5.95
$2.00 for shipping and handling.
Discounts available for quantity orders.
Morning Sun Press Books are available at quantity
discounts when used to promote products or services.

For information please write Premium Marketing Department,
Morning Sun Press, PO Box 413 Lafayette, CA 94549
Phone and Fax (510) 932-1383

Contact with the author through Earth Service Inc.
1800 Avenue of the Stars, # 1190 Los Angeles CA. 90067

Dedicated to my children
Bradford, Jennifer and Laura
and to all the children of the world
Thanks to **Katy Kreitler** for her
patience and thoughtful commentary

Acknowledgments

Kenny Ausebel & Nina Simons for *Seeds of Change*; Ed Begley for walking the talk; Thomas Berry for *Dream of the Earth;* Bishop Frederick Borsch; David Brower for Earth Island Institute; Dr. John Cobb for *The Common Good;* Eduardo Corfino for Gringo Perdido; Peter Cummings for *Talons;* Michael DiGiovanni; The Earth Service Board of Directors; The Episcopal Environmental Coalition and Stewardship Team; Ethan Flad; Vivian Gabehart; Andy Goodman; VP Al Gore for *Earth in the Balance*; Paul Gorman; Tom Hayden for vision; Randy Hayes for Rainforest Action Network; Denis Hayes for Earth Day; Bob Hollman; Jack Howell; Michael Klubock; The Kreitler Family; Andy Lipkis for Treepeople; Brian Little; The Lucas Family; Howard Lyman for Voice for a Viable Future; Rob Maguire; Miriam MacGillis for Genesis Farm; Nina and Mark Merson for Eco Expo; Warren Morse; The Very Rev. James Parks Morton of The Cathedral of St. John the Divine; Randy Olson of the National Geographic; Permanent Charities for Earthwalk; John Quigley; Andy Rasmussen; John Robbins for *Diet for a New America*; Steve Robertson; John Seeley, John Theodore for The Great L.A. Clean-Up; Margaret Trumbull; Fr. John-Alexis Viereck; and the environmental leadership of California, the United States and the world for courage, guidance and commitment against difficult and sometimes overwhelming odds; and to the many other friends who have supported the work of Earth Service Incorporated through the years.

<u>CONTENTS</u>

The Letter "C"

During the past five years a fascination with
the relationship between the "collapse of creation"
and the letter "C" has prompted research by the
author and the creation of this book.

CHALLENGE

Every single child on this Earth needs and deserves a clean and healthy environment, yet a large percentage of the Earth's greatest natural resource lives in places where the environment causes illness and death.   As adults we are morally corrupt if we do not address the **CRIES** of the **CHILDREN** heard 'round the globe.

The most searing indictment of the present generation's legacy is that we have not met the challenge to provide a viable future for the children. This challenge became real to me in the fall of 1990 while driving my 12-year-old daughter to school. We were listening to the news on the radio when we heard that the two environmental initiatives on the California ballot had gone down to defeat. My daughter turned to me and asked:

> "Dad, how do you feel about those two environmental things getting beaten?" "Not very good," I responded.   "So (pause) what are you going to do about it?" I stopped the car and said, "Laura, I am going to do a lot!"

The challenge of my child launched me onto a path that has led me to do whatever it takes to make this planet a healthier place for all of creation. I believe the prime responsibility for changing our current course falls squarely on the shoulders of middle aged persons in power and trust. I personally felt the urgency in my daughter's voice because there is no longer time to mull over the challenge. An instantaneous conversion was required so I called The Rt. Rev. Frederick Borsch, Bishop of the Episcopal Church in Los Angeles, and my boss. We met in his office and I told him that I wished to resign parish ministry and begin a non-profit organization in service to the Earth. I asked for an on-going connection to the church because my priesthood was important to me.

> "What do you want from the church?" asked my Bishop.
> "How about a title?" I responded.
> "What would you like?" continued Bishop Borsch.
> "How about: Minister for the Environment?" I asked.
> "Sounds good to me. Now go into the world and speak, preach and teach about the preservation of creation."

With this second launching, the challenge of the children became clearer to me. I resigned parish ministry in January of 1991 and formed Earth Service Inc. in March of that year. Like an artist with a blank canvas the challenge lay before me to create an organization that would not duplicate what other non-profit groups were doing. There were already strong protectors of the oceans, forests, animals, and every aspect of the complex inter-related eco systems that comprise the Earth. What I could protect was their good work. My voice and actions could support and encourage environmentally conscious persons in all walks of life.

During the year long process of trying to discover how I could be of help and where Earth Service could respond to the challenge a particular image came to mind. I remembered the clock on the wall of the main hallway of the Conference Center in Rio de Janeiro during the historic 1992 UN Conference on Environment and Development, called the Earth Summit. This modern day timepiece captivated me and I sat and watched it tick away for several minutes. The Earth's **CRISIS** was portrayed with a ticking clock, but this was no ordinary clock. The top number was the number of people on the Earth. This number kept going up as fast as I could snap my fingers. The bottom number referred to the amount of arable land still left on the planet. That number was going down almost as fast as the top number was going up. This disturbing image kept flashing in my mind like a neon sign at night. Reflecting upon this visual image caused me to realize I had to become a disciple, a learner all over again. I needed to learn as much as I could, as fast as I could, about the state of the Earth so that I might be able to teach others. I finished reading *Earth in the Balance* by VP Al Gore on my way home from Brazil and his book further encouraged me to pursue becoming an environmental educator.

Earth Service began by asking experts to speak at a monthly environmental roundtable breakfast. For four years I have listened to persons deeply committed to restoration and preservation of our fragile island home. I am now convinced that the bottom line for me is to help foster an environmental ethic as a part of the fabric of everyone's life. In addition, I can help enable people to make the spiritual connection to our sacred home. This must be woven into making practical changes on a daily basis. It is clear that every action we take and every decision we make has a significant environmental connection. .

The urgency to foster conversion to this ethic is reflected in opening and closing this book with a challenge from a 12-year-old. Each chapter asks us to face the reality of a rapidly changing world. Caring adults can no longer be complacent and convinced that someone else will make the changes necessary to ensure their future. The **CHALLENGE** of the **CHILDREN** is to every one. To avoid or deny it is to leave our children adrift on a sinking ship without hope.

CONTEXT

We begin our examination of the environmental crisis and a look at <u>The Earth's Killer C's</u> by establishing a framework for our discussion. The first context to frame our understanding of the story of the Earth today is **CREATION**; the second is a challenge from **CAPTAIN JACQUES-YVES COUSTEAU**; and the third is finding **COMMON GROUND** & **COMMON VALUES** among diverse groups.

The largest context is **CREATION.** Creation is like a giant puzzle with millions of inter-locking pieces. The Earth puzzle is complex, yet each part belongs and has its reason for being. Bruce Babbitt, Interior Secretary of the U.S. remarked that "in every case, that (loss of a) species is a warning light about the decline in productivity of an eco-system." This view of creation is essential to understand. Today we are losing parts of the puzzle on a more systematic and rapid basis than ever before in the 8 billion year history of the planet. The puzzle is literally coming apart. As we lose a species here, another river there, a mountain hillside here, and a pristine lake there, we accelerate the collapse of the whole. Tampering with all segments of creation is now beginning to be seen in context by scientists, environmentalists, and those committed to a sustainable future. Any environmental discussion about trees, plants, animals, air quality, or loss of arable land etc. must be set in the context of the whole. Thus, each <u>Earth Killer C</u> in this book affects the whole of creation and consequently our future.

The second context is the **CHALLENGE** of **CAPTAIN COUSTEAU** that was delivered at the UN Conference on Environment and Development. At that pivotal gathering of over 125 nations, octogenarian and visionary Jacques Cousteau challenged the human family with these words: "I wish at the Rio Conference heads of State and their delegates would realize the urgency of drastic, unconventional decisions. You have an extraordinary opportunity to **CHANGE** the **COURSE** of the world, but only if you decide to challenge the huge problems with radical solutions." His challenge penetrated the hearts and minds of many, including myself, and changed me forever.

Business as usual by nations, individuals, or corporations will continue to plunge our planet into an irreversible course of collapse. What is required is an extreme alteration of direction and that will require an enormous coordinated world wide effort that can be led by the people of the Untied States and the United Nations. Nothing else will matter unless we change the course we are presently on to one that guarantees a healthy, sustainable planet for our children to the seventh generation, that is forever.

The third and perhaps most important of the contextual criteria is to find **COMMON** values and **COMMON** ground from a diversity of opinions. Every one living today knows that the health of the environment is crucial to our well being, but there are vast differences on how to keep it healthy. We also know that most of the human family is motivated by self interest. People tend to protect their own environments. If we love our neighborhood we will become activists if it is threatened.   If we love our children we will safeguard their well being. In the U.S. today the battle cry of Democrats and Republicans alike is to "do something about the violence" inflicted on our children. Part of the increasing spiral of violence nationally and globally is environmentally related. Thus, all must work to find common ground for a complete change of course. Adults are asked to put aside the differences that separate and divide us one from another. Even the 104th Congress is preaching "moral high ground" and cooperation from its hallowed chambers. Our elected leaders say that their common concern is our children's well being. Sustaining a healthy environment with strong legislation to back up the words by creating an environmental ethic as the basis of all decisions will affirm their common concern for the children. Anything less is not acceptable.

This translates into preservation and conservation of our Earthly home's fragile eco systems. The Greek word for home or household is oikos and from this comes our word eco. The place where we may find common ground is for each of us to live out our lives as eco-conservatives; that is dedicated to conserving our home. As a parent of three children I am increasingly radically conservative about this mandate. The open challenge of our children is to conserve the trees, clean the water, preserve the diverse species, and keep our home intact forever. The place for us to begin is in our own backyard with deeds of action, not advertising campaigns to perpetuate our own self interests. My personal contract with America is to **CREATE** a **CONTRACT** with the **CHILDREN** to provide for them a long term healthy environment. There is no finer gift that we can leave our children.

There are three other **CONCEPTS** that have shaped my thinking about the environmental crisis. <u>First of all,</u> and contrary to popular opinion, it is pointless to address environmental concerns with the children of today unless the adults in power are willing to change systems already in place. It is a waste of time, energy and money to teach recycling to kids while industries led by their parents seek to gut key legislation designed to promote recycling and recycled products. It is hypocritical at best to talk about health care for our inner-city children when we allow chemical and vehicle emission violations to go un-enforced.. A sensible solution would be to develop and require environmental education for every adult in every nation in the world. Business schools require a course on ethics today; corporations, academic institutions, communities of faith and trade schools could teach an environmental ethic for adults. Environmental night school could be a fun alternative to the "couch potato blues" and a valid way to accelerate the adults' environmental learning curve. <u>Secondly</u>, the conspiracy of the self labeled "wise use movement" must be clearly understood as un-Judeo-Christian, and un-American. Masquerading as protectors of the environment, a coalition of industry financed groups strategize on a regular basis on means to block sound environmental policy. In essence, it is more accurately described as the "unwise abuse cabal." Trying to undermine the protection of the legacy of wide open spaces, clean air, water and forests threatens to discount the Bill of Rights which guarantees a healthy land in perpetuity. Robbing future generations for the pleasure of the present and calling themselves 'wise' is an affront to the creator and the creation. Protection of self interest at the expense of the common good undermines our ability to provide a viable future for our children and grandchildren. Liberty and freedom for all does not guarantee the right of a few to abuse an inherited gift, and that is exactly what is happening nationwide. <u>And third,</u> there is a profound sense of urgency about conversion to sensible, sustainable, personal and institutional behavior patterns. The Union of Concerned Scientists' Warning to Humanity of Wednesday November 18, 1992 reflects the potential **COLLAPSE** of **CREATION** within my lifetime and that scares me and challenges me to action. I quote:

"We, the undersigned, senior members of the world's scientific community, hereby warn all humanity of what lies ahead. A great change in our stewardship of the Earth and the life on it, is required, if vast human misery is to be avoided and our global home on this planet is not to be irretrievably mutilated. No more than one or a few decades remain before the chance to avert the threats we now confront will be lost and the prospects for humanity immeasurable diminished."

If creation is about to collapse, the human family needs to, in the words of David Brower, "give the planet **CPR.**" Cardiac pulmonary resuscitation can put the breath of life back into the dying person. Today we must recruit leaders willing to breathe new life into the decaying and dying Earth's eco systems.

CAPTAIN COUSTEAU challenged individuals and nations to find "radical solutions". Perhaps in addition to the chaplain's prayer before every session of Congress, The Scientists Warning to Humanity could be read out loud so that Congress would finally understand what they are warning us about. A yearly updated and condensed version could become a part of the on-going information super highway on computer bulletin boards. Even the most ardent environmentalist needs to continually hear the message that we must change course now or else the collapse of creation will continue. Congress is looking for areas of common ground to lead the people of the United States to the promised land of justice, equity and quality of life for all. What can be anymore fundamental than Congress becoming eco-conservative by conserving our precious dwindling resources and restoring that which has already been damaged?

The message eco-conservatives can teach to other adults is to encourage them to do *50 Simple Things To Save The Planet* as a stepping stone to complete conversion; or as Fr. Thomas Berry says, "re-invent what it means to be a human." **COMMITMENT** to **CONVERSION** by every man, woman and child on this planet is the foundation of hope. <u>The Earth's Killer C's</u> offers us the context to change our personal and institutional course of direction. We begin our examination with a presumed benign utensil that is eating away many parts of the natural world.

CHOPSTICKS

Every time we take a bite of chop suey, vegetable tempura or tuna sushi we may be contributing to global warming and·the loss of old growth forests. As strange as this may seem, it is true if we are using of one of the 6 million pairs of **CHOPSTICKS** manufactured every day from Canadian trees. Hundreds of square miles of forest that are not being cleared in a sustainable manner by The **CANADIAN CHOPSTICK** Manufacturing **COMPANY** are disappearing rapidly. Disposable chopsticks are a tiny portion of once magnificent trees that stood proudly on the slopes of countries all across the globe. Malaysian and Indonesian forests have been decimated, and now

Canadian Forests are yielding their aspen trees to the largest waribashi or disposable chopstick factory in the world. This company, like most chopstick manufacturers,   wastes approximately 85% of each tree because the end user wants an unstained chopstick. This is a silent Killer C because few people are aware of the connection between eating food with chopsticks and the environmental crisis.

It is appropriate that we begin a look at <u>The Earth's Killer C's</u> with this seemingly insignificant eating implement. **CHOPSTICKS** represent the cumulative effect of human behavior on the sustainability of the planet. Today, hundreds of millions in the human family use and discard the planet's dwindling resources without a conscious thought.

Six million disposable chopsticks a day, though a staggering figure, is but a pine cone in a forest. Japan alone imports roughly 8 billion pairs of throwaway "little trees" a year. In the City of Los Angeles, which has over 400 restaurants that use chopsticks, two trading companies provide them with 2.8 million pairs of chopsticks a month imported from Indonesia and China. Also, Canadian chopsticks are harvested, sent to Taiwan for finishing and then to Japan for use. It makes no sense that Asian forests are the source of chopsticks for use here in the United States. The net result is wasted trees, wasted energy, and dire world wide consequences. This carnage occurs because we chose an eating utensil that could be replaced with a re-usable or re-cyclable product, or transported to and from restaurants easily. **CHOPSTICKS** could become a symbol of a new vision for a new earth. Converting from the long entrenched practice of using pure, virgin chopsticks would offer hope to a world on the brink of environmental collapse. These two little pieces of wood forming an inexpensive eating utensil reflect how humans treat creation. We use what we want, when we want, and we throw it away where we want. But we can convert with a little knowledge and a lot of will power.

The chopstick Healing C is **CUSTOMER CARRY**. If everyone were to carry their own chopsticks the demand would disappear overnight. Members of families who use chopsticks daily have permanent sets that are washed and reused like others treat a knife and fork. The chopsticks are either made of bamboo, plastic or ceramic. Why not carry these whenever we go out to eat? Perhaps we could design a little sleeve within our coats that could function for the transport of our own eating utensils to Thai, Japanese and Chinese restaurants. It could become cool to carry chopsticks. Also, when visiting an eating establishment that offers chopsticks ask if they are permanent; if not, use a knife and fork and bring your own the next time. Changing the consciousness of someone often results from posing a simple question that had never been asked before.

Recently I was giving a presentation on personal responsibility and the environment in San Clemente California. During the dialogue portion of the evening a young man raised his hand to tell his personal story of purchasing 600 acres of Costa Rican rainforest. "I wish to preserve this so that my grandchildren can enjoy nature in its original form." Everyone applauded and the young man smiled. When I asked if anyone knew how much land had been recently acquired by the two corporations Chugoku and Mitsubishi no one raised their hand. These companies are logging 46,000 square miles of Canadian forests to make up to 9,000,000 pairs of disposable chopsticks a day. Presently they have not reached their full capacity, but the factory can accommodate another 50% growth in manufacturing capability. This is not the logging of a tree farm, but the clearcutting of miles and miles of old growth forest. Place this issue in context for a moment: A diverse multi-national corporation can undo the work of a person's entire lifetime in one day, all to make throwaway eating utensils.

Chopsticks are small inexpensive items that are taken for granted and rarely discussed in any environmental book, magazine or literature, yet they are costing us more than we can calculate. Recognizing the scope of this problem since 1990, **CONSUMER CO-OPERATIVE** Societies in Japan are now **CONDEMNING** the use of **CHOPSTICKS**. They face stiff opposition from manufacturers who say "the chopstick was given by the gods. We think it is something special." Creative anti-chopstick activists are demonstrating how alternatives can be used that will not harm the environment nor alter the taste of the sushi and sashimi.

In our country the issue is certainly more one of ignorance and laziness, rather than an affinity for the product itself. Yet, this little product that we take for granted comes from trees that are more worthy than being used for a disposable item. The next time your mouth waters for the delicacies of the Orient, carry your own chopsticks to the restaurant, and tell a friend why you are carrying your own.   You may start a trend that will guarantee all our grandchildren a healthy, sustainable future, because you are helping save another piece of the Earth puzzle: an entire Canadian old growth forest.

CLEARCUTTING

The Friar Tuck method of clearing trees from forests is bringing whole eco systems into cardiac arrest. Clearing a stand of trees to a state of making the earth bald with a surrounding fringe of forest is the clever way industrial forestry has hidden the truth from the public. However, a few years ago the environmental air force Lighthawk began to photograph timber stands throughout the continental United States, Canada and Alaska from the air. What they discovered made my wife and me cry when we thumbed through the pages of the book *CLEARCUT*, published by Sierra Club and Earth Island Press. Photograph after photograph clearly illustrated that our current method of taking trees is killing the eco systems within which the forests thrive. A single photograph from *CLEARCUT* may be worth a thousand words of concern in a hundred magazines or books.

In the book's introduction, Bill Devall writes: "During the last fifty years -- since the advent of the **CHAINSAW**, tractor, and logging truck, forests have been clearcut from steep mountain slopes in California, Oregon, Washington and British Columbia. Collectively the photographs in this book provide a portrait of the great forests of North America as they cling to life at the end of the twentieth century. They reveal that the forests of North America are suffering from a massive, human-caused, catastrophic event, an epidemic of massive clearcutting."

What is sad is that clearcutting was not broadly practiced until the mid-sixties in our nation and then given legal status in 1976. Sustainable forestry meant selective cutting so that the watershed, the downstream river and forest life, and the soil hosting the trees would remain stable and capable of sustaining life forever. **CLEARCUTTING,** the preferred choice of industrial foresters today, dramatically reduces the chances of an eco system's survival. Where have all the coho salmon gone, one might ask? Ask the clearcutters!

One of the most beautiful stretches of North American wilderness in **CANADA** is around **CLAYOQUOT** Sound. Up until 1993 Canadians were proud to call this their Hawaii. Clayoquot is filled with lush forests along a beautiful coastline dotted with hundreds of islands. Clayoquot Sound is a symbol of the majesty of creation; and yet almost overnight, the Canadian government granted permission for the clearcutting of more than two-thirds of this area, in part for our throwaway telephone books, thereby adding to the already staggering statistic of losing 80 acres of forest a minute worldwide. That number may not seem like a lot, but it adds up quickly to the sum of roughly 40 million acres of rainforest or old growth forest lost every year. Professor Laura Sewell, commenting on her recent trip to Canada, gave us a word picture that tells the story: "I went looking for wilderness and could not find it. I canoed up the whole coast of Canada and could not find wilderness." What this means to children, and not only Canadian children, is significant!

"As we liquidate the ancient forests," writes forest ecologist Chris Maser, "we are redesigning the forests of the future. In fact, we are redesigning the entire world, and we are simultaneously throwing away Nature's blueprints." Today clearcutting the trees from the creator's garden is perhaps the final insulting chapter in our short visit to the Earth. The Tree is the symbol of life itself. Even though trees are a renewable resource, a tree farm is not a valid substitute for a complex eco system that has evolved over eons of time. A "beauty strip of trees" along a highway is not a substitute for a forest either. Both are illusions created to make us believe that we know what is right for the planet. What is right for the Earth is to leave intact those necessary forest ingredients that help sustain all life.

Forests stabilize climate, provide a home for untold species of birds, plants and animals, and give scientists millions of organisms that have the potential to cure human illnesses. Deforestation, on the other hand, releases a mixture of greenhouse gases such as carbon dioxide, methane, nitrous oxide, and disrupts all forest dwellers' habitat.

This is only part of the story. Cutting the cleansing trees at the rate of 115 thousand acres a day while driving millions of additional combustion-engine vehicles that pollute the air presents us with another ticking time bomb. When we lose our forests, we diminish the planet's natural ability to sink the **CARBON** that we produce from burning fossil fuels. I first came across the term carbon sink late one night in Rio reading a paper by Dr. Robert Goodland. He talks about this hidden killer, but only a handful of environmentalists even know what this term means. Scientists are worried that we may be past the point of being able to cleanse the carbon from our atmosphere. The effects of this will be the acceleration of **CLIMATE CHANGE** which will affect the lives of generations unborn. Thus, worldwide forest management is necessary for our long term ability to breathe, and yet we continue to cut 4 billion oxygen-giving trees a year just to make paper. Clearcutting is not a long term viable method of forestry, but a clear short term gain of profit for a few.

In our nation the **CONSTRUCTION** industry accounts for almost 40% of all the wood consumed and much of that comes from the old growth forests of the Pacific Northwest. Conversion to sustainable construction practices is being spearheaded by groups like the EOS Institute who believe the challenge from Christopher Alexander who says 'every increment of construction must be made in such a way as to heal the city.' Conversion to alternative building materials, recycled materials and the elimination of old growth forests as a source for wood in the construction of homes is possible today.

There is another issue related to our forests that makes good fiscal conservatives sick to their stomach. In the 1994 WorldWatch report, the Cascade Holistic Economic Consultants noted that "109 national forests, accounting for 80% of national forest wood output, lost money on their timber sales in 1992. The loss to U.S. Taxpayers was $499 million dollars." We spend millions on the infrastructure to remove the logs and receive so little per acre that losing money is commonplace. If clearcutting our timber looses money and the taxpayers foot the bill, why is there not a revolt?

Also, logging corporations and public land agencies, such as the Forest Service and The Bureau of Land Management, often try to disguise clearcutting by calling it "selective logging", "group selection", "shelterwood cut", or "seed tree cut." Any logging methods that remove most or all of the trees on any given acre, or take the biggest, oldest, healthiest trees in essence create the same ecological destruction as clearcutting. Eco-Conservatives in Congress could save the government millions while saving the forests at the same time if they looked at the numbers and took a plane trip with Lighthawk. Goods produced at a high ecological cost should at the very least not lose money for the nation. There could be an environmental tax on those practices that rob the future from our children. That money could go towards sustaining jobs in the forestry industry in the Pacific Northwest, while maintaining the spotted owl and the coho salmon's habitats at the same time.

Is there hope? Yes, because concerned citizens throughout this nation and the world are giving voice to the voiceless trees and forests by forming **COALITIONS**. These wonderful "tree huggers" know that 2000-year-old redwoods, cyprus, and aspen cannot speak for themselves,. A variety of action-oriented protectors of the trees are rising up in defense. In the introduction to *CLEAR CUT* there is the story of a small group of people who stood up for the Cathedral Forest in the Cascades of Oregon. They write: "We believe that all things are connected; that whatever we do to the Earth, we do to ourselves. If we destroy our remaining wild places, we will ultimately destroy our identity with the Earth; wilderness has value for humankind which no scientist can synthesize, no economist can price, and no technological distraction can replace. The forest, like us, are living things; wilderness should exist intact solely for its own sake; no human justification, rationale, or excuse is needed. We perceive the earth is dying. We pledge ourselves to turning this process around, to stopping the destruction, so that the Earth can become alive, clean, and healthy once again. On behalf of all citizens of the Earth community, we declare the Cathedral Forest -- all that which remains of Oregon's old growth eco-system -- an inviolable wilderness for all time." Oregonians are proud of their long standing good forestry management practices. Clearcutting is an insult to their way of life.

A Cousteau thought might be that whenever we say the pledge of allegiance to the flag of the United States we make a commitment to a cause on behalf of those parts of creation that need our passion and voice. Sitting idly by while the forests literally burn or are torn up by the roots is a sin of omission that the planet can not tolerate.

Clearcutting takes more than the forest and the eco system, because it robs us of our sense of awe, majesty, and our capacity for wonder. I have stood in the redwood cathedrals of California, hiked in the Himalayas of India and the rainforests of Brazil and Costa Rica. I have explored the jungle forests of Guatemala. I have been surrounded by the pine wilderness of the Pacific Northwest and I have walked, on many occasions, through the most diverse private forest in New England, the Edward C. Childs' Great Mountain Forest in Connecticut. Never does one feel closer to the Creator than when privileged to walk through one of the Earth's magnificent forest cathedrals around the globe. It is clear that **CLEARCUTTING** is a **CRIME** against **CREATION,** yet we keep clearing more and more trees from the land, and the hillsides, often just to graze cattle.

COWS

"Killer **COWS**": Sounds like the title of a grade B science fiction movie. But, this screenplay is real. The devastating effects of a beef-centered diet and an expanding cattle industry are being felt worldwide. The total weight mass of cattle on Earth exceeds that of human beings. Estimates range that between 1.6 billion and 3.3 billion cows inhabit and ravage our finite planet, and we continue to clear forests to make room for more and more.

The raising of "Old Bessie" out on the back pasture among the flowers and spreading oaks is an image of a by-gone era. Cows live primarily in feedlots for their entire life. They are a commodity bought and sold to the highest bidders. These enormous animals produce roughly 20 billion tons of waste every day. Dr. George Borgstrom estimates that U.S. livestock account for 5 times more harmful organic pollutants than all 250 million of us Americans combined. Animal waste pollutes streams, lakes and the coastal eco systems. The amount of waste created by a single 10,000 head feedlot is equivalent to the waste of a city of 110,000 people. That is a substantial pile of (please pardon the expression) **COW CRAP**.

The devastating impact does not end on the ground, but extends into the atmosphere as well. Cows produce methane from both ends of their bodies to the tune of about 100 million tons a year. Methane gas contributes to ozone depletion and global warming. In other words, cows affect the weather. Thus, **CLIMATE CHANGE,** a potentially catastrophic environmental problem, is a direct result of the diet choices we make on a daily basis. When we choose to eat hamburgers and steaks we contribute to one of the causes of changing weather patterns. Also, we now know that ozone depletion leads to an increased incidence of cancer. One of the Healing C persons in our global community is **Helen Caldicott** of The Physicians for Social Responsibility. Dr. Caldicott recently told a stunned audience in Queensland, Australia that 100% of the men of Queensland over the age of 65 had skin cancer. Patching up the ozone hole is not like repairing a pair of worn jeans, but the consequences of not doing so is staggering to comprehend. What can a global citizen do to help close the hole and prevent this increasing cancer catastrophe? One easy step is to examine our diet. What we put on the end of our fork is a choice we exercise daily. The "close the hole campaign" is simple: switch to eating more vegetables and grains on a regular basis.

Yet, clearing land for cows is big business worldwide and ranchers, farmers and governments are cashing in on cows. Fast food outlets are expanding and with them the demand for beef. Forests are being cut, pastures are being fenced, and the world wide loss of topsoil can be measured in billions of tons. Even in parts of the fertile crescent of the U.S. we have lost as much as one-third of our topsoil. More than 80% of topsoil loss is directly attributed to the grazing of livestock. Soil gone today does not get replaced for hundreds or even thousands of years. At the UN Conference in Brazil the Minister for Energy from Morocco warned "that the alarm throughout the world can not leave us indifferent." The concern he expressed was over the loss of the world's nations' ability to grow food. The soils of the Earth's gardens are literally washing into the sea, in part because cattle are leaving their footprints on the planet, and they do not walk lightly. The countries of Haiti and Madagascar, where extensive deforestation has occurred, are living proof of what can happen worldwide: when there are no more trees, there is no more soil and then there is no more food.

The many problems associated with the cattle industry prompted Paul Hawken, in his book *The Ecology of Commerce*, to write that "when cattle ranchers clear rain forests to raise beef to sell to fast food chains that make hamburgers to sell to Americans who have the highest rate of heart disease in the world, we can easily say that business is no longer developing the world. We (business) has become its predator." Laws favoring and protecting the cattle industry may contribute to the literal eating away of the basic elements that sustain life. For example the general market price of a hamburger is around $1.00. The ecological price, according to Alan Durning, who wrote *How Much is Enough*, is close to $200 per hamburger. We are well aware that our air, water, and soil are under constant attack from cows, yet this dilemma receives far less attention and less protection than the predators, and the cost for the damage is not even factored into the price of the product.

Howard Lyman, a fourth-generation cattle rancher from Montana converted to a non-meat diet because he finally recognized the damage being done by his business to the environment, and to the health of his children. Lyman owned 12,000 acres of grain production, a 5,000 head-feedlot, and 1,000 head of range cattle. He became a wealthy man and remained a staunch believer in the lessons he learned in college about how to run a chemically dependent cattle ranch.

But Howard Lyman had a conversion experience. His own illness, and the death of many of his buddies from cancer, convinced him that he should change his way of living and working. He sold his ranch and today travels the world convincing others to convert to organic vegetable farming and to consume no dairy products at all. He talks about the damage cattle cause to our nation's water supply, the inefficient use of energy to produce a pound of beef , the amount of grain fed to non-humans, (1992 statistic: 635 million tons of grain to feed livestock and poultry), and the inordinate amount of pesticides and chemicals used in the industry. I really took notice, however, when he spoke about how we are now feeding **COWS** to **COWS** and the potentially dire consequences of this practice. In other words, natural herbivores are being made into carnivores. Why? Because 125,000 cows a year are falling over dead and we have to do something with this enormous mass of meat. Howard saves the worst for last. He astounds his audience when he talks about the increasing prevalence of "mad cow" disease. "Mad cow" disease can be spread through both the animal and human populations and there is no known antidote. This is such a virulent disease that pathologists will not even examine the tissues of these dead animals, and our government is already banning sausage casings and animal products from certain countries. This is scary stuff and should cause us to look carefully at the "sacred cow" industry.

Noted authors Jeremy Rifkin and John Robbins, in their best selling books *Beyond Beef* and *Diet for a New America*, detail the value of **CATTLE CONVERSION** and switching to a non-meat centered diet. The number and diversity of statistics they offer can convince even the staunchest beef lover that conversion is necessary for the long term sustainability of the planet. For example; eating one double cheese burger equates to the loss of another 55 square feet of rainforest that had to be cut to pasture the cattle. Robbins and Rifkin have convinced skeptics that for their own self interest and the well-being of their children, conversion is the smart choice. Their books also provide good information for common cocktail party conversation. This is an arena where environmental education has to take place in many of our communities. Personal one-on-one chats about the environment with solid facts can make a difference. One of my favorite topics is dirt because it opens up the dialogue in many different directions. I talk about composting as the remedy for the waste problem created by cows and the loss of topsoil. We are losing topsoil at 18 times the replacement level we need. Why not create thousands of **COMPOSTING CENTERS** tha turn our waste into the treasure of new soil? "Soil is the diversity which gives life," commented David Brower. Composting is nature's way of utilizing cow, yard, food and organic waste. This topic usually leads to a discussion of organic gardening and diet. The next step is to explain why our family no longer eats meat and is eliminating dairy products altogether. Usually one or two health examples, and one environmental example is enough to convince our friends that we are not crazy. Thoughtful dialogue is a good way to encourage a friends conversion to new environmentally sound diet choices and behavior.

In California, a major source of the nation's food supply, it is estimated that growing lettuce, tomatoes and potatoes requires approximately 25 gallons of water to produce each pound of food. A pound of eggs requires 130 gallons, a pound of chicken 800, a pound of pork 1600, and a pound of beef upwards of 5200 gallons of water! Yet Californians are asked to conserve water by shortening showers, not washing cars at home, and planting drought tolerant gardens. Nine years ago our 7-year-old, during a drought, remarked that if she gave up eating hamburgers she could take all the tubs and showers she wanted. Translated into water use that means it takes someone showering 5 times per week, 52 weeks a year, 5 minutes per shower, with a flow rate of 4 gallons per minute, to use up the same amount of water used to produce one, that is one pound of beef. This fact alone is reason enough to convert to eating lower on the food chain and giving up beef totally. Thus, our family realizes that not eating beef saves water, soil and the protective ozone layer.

However, it is unrealistic to expect that conversion to a **COW**-free **CULTURE** can happen overnight. Resistance to change is common, yet there are persons willing to hear about alternative patterns of behavior. There are ranchers like Bill Barnard who turned his Cathedral Bluffs Colorado grazing allotment from the federal government into a model of range stewardship. Commenting in the May 1994 issue of National Wildlife Magazine, and reflecting a growing sentiment, Barnard said "I've been a cow man all my life, and it's my feeling that you better get around and take care of your cows; if we don't start taking care of our public lands, we'll get kicked off." Restoring degraded pasture land is the responsibility of the ranchers of America. Many are beginning to change their practices and the public can help by supporting dairy and cattle ranchers who are producing without pesticides, or using growth hormones, and who treat their animals in a humane way.

All of the above are positive steps, but it is my personal belief that eventually all cows must give back the land to vegetables and grains as we become vegetarians. This word comes from the Greek "vegetas" meaning "full of the breath of life." The human family, especially the children, can thrive on a meatless and dairy free diet. Converting our grain growing to feeding the children instead of the cattle is a positive first step, but we can not stop here.

CARS

Place any living creature in a closed system and turn on even a new combustion engine and death is the consequence. Our Earthly atmosphere is a closed system about 10 miles high. Thus, 5.6 billion people's lives are threatened because the burning of fossil fuels, primarily in cars, busses and trucks (**C-B-T's**) is polluting the air. A new **CAR** arrives on the world's **CEMENT** and **CONCRETE** network of roadways and bridges every second of every day, thereby threatening the long term sustainability of the living world.

In our society, we demand that killers be jailed and many believe they should be killed; but we have licensed 140,000,000 "killer" cars in the U.S. alone. This ever expanding squadron which includes trucks and busses, has an accomplice in the destruction of the environment. The enormous **CEMENT & CONCRETE** infrastructure builds the roads and bridges which allow us to drive a trillion miles a year. We are paving more and more of the world to move ever expanding populations in vehicles that pollute the air necessary for life. Three years ago on Earth Day, John Quigley, my partner in creating The Great L.A. Clean-Up and I went aloft in the Goodyear Blimp over Los Angeles with a reporter from the local CBS affiliate. He asked, "What do you guys see when you look down?" Our candid comment was, "asphalt and concrete!

In L.A., the car capital of the world, the Century Freeway was just completed at the cost of $100,000,000 dollars a mile, and L.A. is where an average freeway mile costs around $10,000,000. In 1991 Californians pumped 35 million gallons of gas into their guzzlers every day to drive on these freeways at 30 miles an hour. By the year 2010, a current commute that takes 45 minutes will increase to more than 2 hours. Californians are not alone. **COMMUTING** has become a Killer with a small c in many cities around the globe. I recently returned from the beautiful country of Guatemala where the pollution from their vehicles is staggering. Steady streams of thick black exhaust pour out from tailpipes on their diesel trucks and busses as they transport their people around the country. Commuting in congested corridors kills our spirits, diminishes our family time, and affects our health. Work hours by the millions are lost to freeway congestion. In addition, in my backyard, we Los Angelenos live with air that is called "lethal." Even with buy back programs, stringent clean air standards, cleaner burning gasoline and engines our smog problems continue because of the sheer numbers of **C-B-T's** on our roads and freeways.

However, things are getting better in L.A. thanks to the cooperative efforts of many different groups and individuals.

But, other urban centers of the world are becoming increasingly polluted and congested; and this is where the population growth is occurring. In several cities the engines of **C-B-T's** have created an environment of no return. Santiago, Chili, Mexico City, Mexico, and Hong Kong are three places where the air quality is reducing the lung capacity of entire generations of children. Edward Abbey writes that "perpetual growth is the creed of the cancer cell." Unchecked growth of our cities and the introduction of the vehicles that support the infrastructure are literally creating cancer corridors in cities. In East Los Angeles autopsies were performed on 18 teenagers who had lost their lives in gang warfare. All eighteen had significant lung damage. Cars emit carbon monoxide, nitrogen oxides, hydrocarbons, and ozone. All of these chemicals become part of our daily inhaling and exhaling patterns, especially for our poor. Most of the critical damage is done to people of color, urban city dwellers, and children whose lungs are growing. Critics of this urban crisis for the poor and disadvantaged look to the automobile and diesel trucks as a key to the solution. Cleaner **C-B-T's** equal healthier inner-city kids.

In 1990, Toyota labeled Thomas Berry, author of *Dream of the Earth* the "devil incarnate" because he suggested that we let the car infrastructure collapse. His answer to the crisis was to not build more freeways or repair bridges, and let the land reclaim green space. Today there is an increasing sentiment and commitment to begin de-paving America. The Alliance for a Paving Moratorium publishes a newsletter called the Auto Free Times in which they offer up a variety of Healing C's for the planet. Clean urban environments are critical to healing the planet and there are now many conversion options that can help clean up the air and reduce the car congestion and contamination.

Conversion to sensible technologies already available is a good first step.   Catalytic converters, carbon flo technologies, ceramic engines, combustion efficiency management plans, adherence to Corporate Average Fuel Economy requirements, and cleaner burning fuels all make a difference. The first goal is to have cleaner, more fuel efficient cars become a larger percentage of the cars on the road.The second goal is to get out of cars and get rid of cars, beginning in our cities. Widespread conversion to **COMMUTER CARPOOLS** relieves congestion.   This sentiment is now being shared by a wider spectrum of the American public. Corporations and businesses are concerned that their workers are losing valuable time on the congested freeways. Strong alliances are forming to promote mass transit, car and van pools, ride share days and weeks, as well as cleaner vehicles. Sick workers are non-productive workers and driving through the daily dose of ground level ozone smog is affecting the health of thousands. Many organizations like The Coalition For Clean Air in California are promoting conversion to electric vehicles, or alternative fuel vehicles. Also, scrap schemes are being funded that buy back clunkers that are a major source of pollution in our cities. Diesel engines in particular are being targeted for conversion to alternative fuel consumption. L.A. leads in the conversion race because non- profit groups are working closely with the California EPA, the California Air Resources Board, the South Coast Air Quality Management District, and the Mayor's Office. L.A. is a leader in fostering common ground and compromise among those diverse agencies working for a clean Los Angeles. Cities worldwide should take note to convert now before the air quality reaches crisis conditions. Cities can learn from reflecting upon the L.A. experience.

Primarily because of **C-B-T's** there is about 3 times more carbon dioxide in the atmosphere today than there was prior to the Industrial Revolution. This stuff has to go someplace or else we will all get sick. One of the major gifts of nature, and an asset to the makers of cars, the producers of oil, and the cement companies are the Earth's natural cleansers. The important trinity of plants, plankton and soil absorb the bad stuff that our combustion engines release. Cars are like giant vacuum cleaners. They suck in oxygen to such an extent that a single car traveling at 55 miles per hour for 30 miles consumes the equivalent amount of oxygen as 30,000 people breath in an hour. In comes air and out goes pollution that must be absorbed. Thus, **CREATION** itself is a Healing C because it takes in and processes our carbon waste. Yet, we keep cutting trees and making cars.

As oil companies develop cleaner burning fuels they should also be acknowledging that millions of stands of trees can help them clean up the air naturally. The Healing C of cooperation between leaders of corporations and environmental groups would help us all enjoy cleaner air. In Los Angeles County, the Chairman of Arco and the Executive Director of The Coalition for Clean Air's offices are separated by 20 miles of concrete that carry more cars than any other freeway in the world. These leaders should serve on each other's board of directors. That level of cooperation could begin to make a real difference in air quality because they share a common concern about their own city.

In America we are beginning to make the connection that the present situation is untenable. Fascination with the **CAR CULTURE** is waning as we realize the health and personal convenience liabilities associated with our total car dependence. However, our car companies seem eager to convert bicycle friendly countries to car cultures overnight. Wholesale conversion to cars from bikes may accelerate disaster, yet quoting from the Fall 1994 issue of WorldWatch we read that L.A.'s nightmare is presently being exported to **CHINA**: "In 1994 alone, 10,000 miles of new highways are taking a toll as the Chinese construction crews pave their way across wheat and rice fields." India's and China's populations are exploding at almost 3 million potential drivers a month. If good economics means our car companies are allowed to sell "unclean" cars and trucks abroad we automatically contribute to the rate of our own demise. Even selling clean cars to countries that do not have the infrastructure to keep them running clean is asking for disaster. Thus, in an odd way a Healing C is the country of Cuba that has gone from 70 thousand in 1990 to 700 thousand bicycles in 1992. Bicycle cultures are keeping the planet healthy.

CONVERSION in the U.S. to cleaner vehicles and alternative transportation is happening, however, often too slowly for our urban children. There needs to be a radical shift in thinking about the urban environment itself. People are seriously talking about changing from freeways to Greenways. Freeways are **CONCRETE** and **CEMENT CORRIDORS** that exist in sufficient numbers. Greenways are tree lined pedestrian walkways and carefully thought-out bike ways that allow someone to cross the urban environment comfortably. Providence, Rhode Island, is revitalizing its spirit and its downtown by creating a Greenway along the Woonasquatucket River. This Greenway is bringing hope to a city where freeways have cut apart neighborhoods.

An unusual healing C for the urban crisis is the cooperation of colleges and cities. Colleges can provide technical, architectural, and creative help to cities that are constantly short of dollars for conversion projects. Also, **COLLEGE CAMPUSES** are noted for being bicycle friendly and students will support Greenways. Spreading that way of life into the surrounding cities and towns gives many people an alternative transportation system that is healthy and non-polluting. Converting back to a bike culture with additional Greenways is smart thinking and enhances the wisdom of the Talmud which reminds us that it is 'forbidden to live in a town that does not have a green garden.' Inner-city tree lined streets and Greenways will increase the health and pride of millions of people. Urban re-forestation projects are springing up all over the country with the guidance of city planners, environmental organizations, and community neighborhood coalitions.

From the East to the West the de-paving of America is providing a new vision of environmental health. There is perhaps no more poignant a symbol than what has happened around **Cuyhoga, Ohio.** In the '60s the Cuyhoga River caught fire because it was so heavily polluted with oil. Today, 20 miles of an 87-mile Greenway corridor have been refurbished as part of the Cuyhoga Valley National Recreation area and the river is clean.

The federal government is beginning to fund projects like this through the Intermodal Surface Transportation Efficiency Act. Linking urban, suburban and rural areas via bike and pedestrian pathways creates a new climate of hope throughout our country. The extensive bikeways of **Cape Cod** have become a source of enjoyment for thousands and across the ocean in **Copenhagen** they have built a network of 200 miles of bicycle paths. The **Chattanooga** Revision 2000 program is tying in some of its wealthiest and poorest neighborhoods with Greenways. These programs and many more like them across the nation and globe represent the kind of creative thinking that the environmental community and local governments have brought to bear for the common good. The job before us is enormous. Neighborhood after neighborhood must consider the conversion options available and then form coalitions for change.

The United States population represents 5% of the world's total, yet we drive 35% of the world's cars. As large consumers of gasoline we contribute more than our fair share to the over-all pollution problem. 99.9% of the US population averages 25 times the emission level allowable per person to keep the air healthy. Scientists say that we need a 60% reduction in global CO_2 emissions to halt the dominant greenhouse or global warming gas. Until we choose to drive zero emission vehicles or create bicycle friendly corridors, we all are contributing to our children's lung disease and the increase in our own insurance premiums.

Cities can become almost car free if the will of the people dictates conversion. We can look to places like **Curiteba,** Brazil for leadership in attempting to make their city of 2,000,000 people healthier by having an extensive, efficient mass transit system. Human powered transportation is feasible in many of the major urban areas of the world, and clean burning busses and electric trams can provide the necessary mass movement of people required during working hours.

We can also endorse and use commuter bike lanes. Presently there are 4 million bicycle commuters in the U.S. These men and women are environmental hero's. Their choice to get out cars and busses benefits the health of everyone, reduces congestion and saves them money.

Cash is always seen as a Healing C. Freeing cash for environmental watchdogs like the Labor Strategy Center in Los Angeles, and other groups which challenge the car culture is cash well spent. Many environmental activists are modern day Florence Nightingales because they are trying to preserve the health of our children by forcing conversion to products and behavior patterns that keep the air clean.

Also, if we acknowledge the true cost of operating a car we should be willing to pay $5.00 per gallon at the pump. This might hurt us in the pocketbook, especially if we are on a fixed income, but the added revenue to the state treasury's could fund programs to keep us healthy. A radical thought might be to create a non-profit energy company that charges $3-5 a gallon for gasoline. The profits could go towards creating an infrastructure of Greenways and electric vehicle fleets. Many Americans would be willing to pay more for gas if they knew that the profits were not for salaries and bonuses (the CEO of UNOCAL made 16 million dollars in 1993), but for the non-profit development of alternative transportation systems. Creative **COST CONVERSION** means simply paying more at the pump today so that we minimize the environmental and health costs tomorrow. Getting out of our cars and into the convenience of mass transit will take time, but we all may require oxygen tanks next to our gas tanks throughout our cities if we fail to convert to available technologies today.

The final and perhaps best idea to help us convert is to offer the driving public a financial incentive to buy fuel efficient cars. The idea is simple. "When you buy a car, you get a rebate if it is efficient, or pay a fee if it's a guzzler," explains the Rocky Mountain Institute. This is a feebate and could give automakers the added incentive to bring to market their experimental fuel efficient vehicles that can obtain 60-90 miles to the gallon of gas. We could also adopt the Singapore plan which charges drivers to use roads during peak commuter hours.

The figures and consequences of our world-wide dependency on fossil fuel to power our **C-B-Ts** are staggering. From 1978 to 1990 there were 20 oil spills larger than the Exxon Valdez. The total amount of crude oil washed into our global water systems was a staggering 750 million gallons. The aggregate number of vehicles worldwide is around 500 million and the total yearly output of new vehicles is around 35 million, or roughly one car or truck for every three children born. The number of persons per car in the U.S. is under 2. On the opposite extreme, the number of persons per car in China is 1055. Imagine what **CHINA'S** air will be like in the year 2020 when the adults are driving cars and the bicycles are left to the children! Ultimately, the private car with the combustion engine, unless it becomes zero emission capable, must go extinct before it forces the human family to go extinct.

Our children know the reality of diminishing air quality and that is why several young people's environmental organizations combined to present over 100,000 petitions to President George Bush in the 1992 **C02 CHALLENGE**. The children's environmental coalition called Earth Force demanded a reduction in the pollution we are creating by driving C-B-Ts. These were our own children from across the country. If the car culture does not heed their cries, who will they listen to?

CORPORATIONS

There are also others who must listen to the children. The persons who captain the corporate ships today are most often middle-aged men with children. Thus, the directive "lead or get out of the way" is appropriately directed at these **CEOs** (chief executive officers). Randall Hayes, Executive Director of the Rainforest Action Network, challenges the executives: "If you want to discuss this further there are a number of wise people who have something to offer to help you towards the environmental u - turn."

Thankfully, the corporate u-turn or about face is underway in many parts of the world. Industries in countries like Japan and Germany understand the economic and environmental connection and are proceeding rapidly with new technologies and new policies. In our country, this new ethic is well represented in the following statement from the Bechtel Corporation in Policy #107: "Environmental accountability is at the heart of our decision making process." This attitude, if and when put into practice, becomes an important ingredient in the resolution of the environmental crisis. Restoration and preservation of the planets eco systems ultimately rests in the hands of those who control production. Thus, at the heart of converting corporations to an environmental ethic is the heart of the **CEO** who runs the show. James Conlon writes: "trusting the people is the indispensable precondition for revolutionary change." I want to trust the CEOs, and I believe if you **CHANGE** the **CHIEF** executive officer you **CAN CHANGE** the **CORPORATION.**

If the resolution of the environmental crisis were only that simple. Recognizing that trying to change anyone is a tough job, the Healing C groups Campaign For Cleaner Corporations and The Council on Economic Priorities publish a list of the worst corporate environmental offenders. I will not name their names here, but perhaps a list of these corporations should be included in the nightly news reports. Instead of 3 minutes of baseball scores, why not have 3 minutes of the corporation environmental compliance scores? These are the scores that are really important to the lives of our children. No one likes to see their name tarnished, especially with regard to their "good corporate citizen" image. Yet, if this TV tactic is not practical in gaining conversion or compliance, posting the executives' names on the locker room door of the local country club might be effective. If none of this works, maybe changing the heart of the CEO could be done by accessing the conscience in his or her family structure. That conscience may be the spouse, but most likely the children.

If we ask the world's children, especially those children whose parents are corporate leaders, to express their concerns in a "dinner table campaign" we will greatly increase the chance of conversion to sustainable practices. **CHILDREN'S CONCERNS** to **CORPORATE CHIEFS** addresses a fundamental axiom that bears repeating as often as the words to the National Anthem. Human beings protect what and who they love. CEOs are no different than anyone else. Their children and grandchildren matter to them. A children's campaign inviting them to embrace the **CERES** Principles, the modern day company 10 commandments would be an opportunity for CEOs to show that they do listen to the voices of their own children. The 10 Ceres Principles in capsulated form are: **1.** Protection of the biosphere. **2.** Sustainable use of natural resources **3.** Reduction and disposal of waste **4.** Energy conservation   **5.** Risk reduction   **6.** Safe products, goods and services **7.** Environmental restoration **8.** Informing the general public **9.** Audits and reports **10.** Management commitment. One has to have a heart made of stone not to heed the cries of the world's children. It is even harder to avoid the challenges of conscience of one's own children.

Changing the heart's of the board's of directors of major corporations may not be as easy. A radical decision might be that every board could nominate in a public forum a conscientious objector who is willing to challenge the board's policies without fear of criticism or condemnation. The former VP of Bethlehem Steel Corporation wrote, "I believe there needs to be more intentional checking of the decision making process in our corporations." Curtis Moore, author of *Green Gold* commented to me "because corporations are increasingly global they are tough to regulate." Therefore, it would make sense that the Director of the Rainforest Action Network should serve as the **CORPORATE CONSCIENCE** of the huge multi-faceted Mitsubishi Corporation. Or closer to home, the President of the National Wildlife Federation, could sit on the board of MAXXAM Corporation which is clearcutting forests in the Pacific Northwest. The President of the National Audubon Society would make an excellent member of the giant paper company Louisiana Pacific's Board of Directors. The CEO of The Nature Conservancy would be appropriately placed on Texaco's Board, where he/she could talk directly with them about the effects of drilling in the rainforests. In essence, a clear directive for an enlightened minority is not fighting the majority, but showing them how to care for creation. Everyone must be open to new learnings and dialogue opens doors, even to normally off limits board rooms.

Joseph N. Towle, in *Ethics and Standards in American Business* encourages corporations to "behave so that you can justify your action before a technically competent and impartial board of inquiry." By targeting the top 500 multi-national corporations, who control roughly 70% of the world trade and 80% of foreign investment, we maximize the potential for systemic change. Another radical idea might be in addition to adopting species, highways, and beaches; adopt a chief executive officer. A diverse group of 25 environmental organizations could host a conference with 25 top corporation executives and spend three days discovering mutual common ground. Active listening to the concerns of CEO's by environmental activists and vice versa is a step on the road to conversion. This could foster long term cooperation and avoid capitulation by either side. It is important to create a win-win situation without hiding from the truth, covering up past problems, or ignoring the reality of **CORPORATE COMPLICITY** in the environmental crisis. It seems impossible to expect self regulation today. Therefore, a program where business informs the activist and the environmentalist informs the corporate leader would promote the best interests of the children of the planet. It might also avoid extensive and expensive lawsuits filed by the

Environmental Defense Fund, the Natural Resources Defense Counsel, or the Sierra Club Legal Defense Fund.

There is a caveat or warning, however, because corporate green washing or willfully trying to deceive the public about a company's environmental commitment and ethic is an issue today. Alliances need to be carefully structured. No money should exchange hands for the privilege of this mutually beneficial connection. Money would contaminate the process of listening, learning, and finding alternative solutions to bad or poorly thought out environmental policies.

The bottom line of corporations can no longer be exorbitant profit at the expense of the environment. The **CODES** of **CONDUCT** of corporations, especially multi- and trans-national corporations, based upon the **CERES** (Coalition for Environmentally Responsible Economies) Principles is a strong first step toward the conversion required for a sustainable future. The biggest corporations should be targeted first. A 1992 United Nations study found that 50% of all greenhouse emissions, half the oil production, most CFCs, nearly all the cars, and a large portion of the electricity was generated from transnational corporations. However, the **CORPORATE CULTURE** is as hard to change as the Exxon Valdez was to stop. The engines of that ocean going giant were shut down, yet the fully loaded oil tanker required five miles to stop. Applying the breaks today to businesses that do not have an environmental ethic hopefully will produce results by the year 2000. The cooperation of all the transnational corporations and their chief executives is crucial.

The **CORPORATE CONSCIENCE** that is shifting is doing so because leaders care about their children, and ultimately the health of the planet. Inter-generational learning is at the heart of the new paradigm shift to an ecological age. That learning has to begin with adults in power and their children. It is also now time to form coalitions within corporate board rooms to accelerate the change. If this does not happen immediately, product boycotts, non violent confrontation and lawsuits will remain the order of the day. The pace of change to sustainable policies needs to be accelerated, in part because of the sheer volume of people on the planet today.

The leadership of persons in control of the major corporations in the world today should be judged by their environmental ethic and not the "bottom line" of profit. Stockholders should ask for a clear contract with the children that ensures them a sustainable future.

CONCEPTION

Every day, 365 days a year, in a span of time as quick as a heartbeat, two more human beings join their 5.6 billion brothers and sisters around the globe. That equates to 180 births every single minute, 10,800 per hour, 259,200 per day, 1,807,400 per week or 93 million, 984 thousand, 800 new consumers of the world's resources every year. A new China is born every decade. It is obvious that there are too many **CONCEPTIONS** and not enough **CONTRACEPTION CLINICS** for our finite planet. It is clear that runaway population growth coupled with consumption and the widening rich-poor gap spells eventual environmental **C O L L A P S E**.

The statistics are almost too overwhelming to fully comprehend. But the population crisis becomes tragic and personal when we discover what is happening to children in Columbia, South America. On the average, six children are murdered every day on the streets by the "authorities." This form of social cleansing relates directly to the world's increasingly callous regard for the children already here. This story from Columbia may be unusual, but congested cities with expanding populations are becoming centers of violence, graffiti and environmental decay. The natural infrastructure that supports the human family is over burdened because the family has grown too large. The United Nations medium population projection now shows world population reaching 8.9 billion people by the year 2030. The question is, will we even be able to sustain life as we know it until then? The answer is already no in many parts of the world and that should be the concern of the entire human family.

Conceptions worldwide slowed appreciably during the 1970s, but today have reached the point where the carrying capacity of our island home is being critically tested. The definition offered by Sandra Postel, in the 1994 State of the World report, is that "**CARRYING CAPACITY** is the largest number of any given species that a habitat can support indefinitely." The word indefinitely is key. Today, because of birth projections, greater poverty, declining food security, a widening of the rich-poor gap, and a diminishing resource base very few scientists believe our eco systems can sustain human life indefinitely.

Perhaps of even greater significance is that the human family has reached the limits of a **CARING CAPACITY** worldwide. There are presently too many children that are neglected and not cared for adequately today. I remember watching the marauding tens of thousands of children descend from the favellas of Rio to the posh beaches of the Copacabana. I also remember being horror struck as the police treated the children as if they were a hoard of runaway vermin. Sadly, it is estimated that 40,000 children die daily of malnourishment, environmentally caused diseases, and the absence of the caring capacity of cities and governments. How we treat our children is a direct indictment of our conception policies worldwide. Irrespective of the scientific debate about how many people can be carried by our planet today, it is clear we cannot care for those already here. It is immoral to subject children unborn to the same fate. We are birthing children into a painful existence that is an environmental coffin. Comprehensive birth control can't be optional in a world where millions of children go to bed hungry every night. Control is a moral responsibility of a caring population from the Pope to the pastor to the plumber. **CARING CONTRACEPTION CLINICS** are essential worldwide.

The call here is for extremely radical solutions. We are robbing the future by archaic, outdated global policies controlled primarily by short-sighted, self-serving men. The clear call of conscience today is for a female Pope and a female head of the United Nations who will heed the cries of the children and their mothers.

Regarding this issue, I am convinced that only women can lead us from darkness to light. When women are given the opportunity to express their concerns about conception and contraception we hear words like this: "We (the women of the world) demand universally available, woman centered, and woman managed comprehensive reproductive health care, including pre-natal and post natal care, safe and legal voluntary contraceptives and abortions, sex education and information for boys and girls, and programs that educate men on male methods of contraception and on their personal responsibilities." This overtly applauded statement represented the consensus thinking of the women at the 1992 Rio Conference in Brazil. Two years later, Nafis Sadik, the leader of the **CAIRO CONFERENCE** on **CONCEPTION** spoke for millions when she said: "Today's women want smaller families, yet there are 300 million women in developing countries who have no access to modern family planning services. Population stabilization emerges naturally from education, health and welfare policies." Men lead or get out of the way or give the women voice and power to solve the crisis of too many conceptions.

Complaining about the crisis does not solve the problem, and creating more conferences does not solve the problem. The human family must focus on the Healing C's of **CONSCIOUS CHOICE** and **CONTRACEPTION**. If we listen to our own inner voice and examine in our heart the pictures of the children from Somalia, Rwanda or Haiti, then we become willing to shift our priorities and convert to sensible birth control policies. **CONVERSION** to a **CONTRACEPTION** economy rather than a military economy will hasten the return to a healthy environment for billions of the world's children. Providing the equivalent amount of dollars we spend on weapons of destruction for implements of contraception will stabilize the world at a greatly reduced cost.

The next generation must appeal to the **CONSCIENCE** of **CONGRESS** to fund family planning clinics in this country and abroad. The challenge of the children can also be directed at the policies of the **CATHOLIC CHURCH** and other communities of faith that do not promote population control. What kind of authoritative document will be required for a shift in thinking by Congress and the Catholic Church? They have the power to effect change for the common good by instituting responsible family planning agendas with financial teeth. Perhaps these words by the United States National Academy of Science and the Royal Society of London will lead to the inner conversion of adult men and women in power in Congress and the Church: "If current predictions of population growth prove accurate and patterns of human activity on the planet remain unchanged, science and technology may not be able to prevent either irreversible degradation of the environment or continued poverty for much of the world." If that does not convince them to change their policies, this sobering statement from population expert Robert Gillespie might: "It will require one child per family for the entire human race for the next forty years just to stay where we are." And if that fails to inspire them to change, I invite them to take a walk with me through the streets of Rio de Janeiro, or probably any major city in the world to hear the **CRIES** of the **CHILDREN.**

My visit to Brazil and the UN Conference fixed in my mind forever the need for dramatically limiting births over the next 25 years. While walking through one of the many tunnels of Rio I happened upon a cardboard shanty that was home to a 12-14 year old girl and her dog. The tunnel was so fogged in with smog that I had placed a handkerchief over my nose and mouth as I walked under the city. As I passed by in almost total darkness, I noticed this child held an infant of a few months to her breast. The **CARING CAPACITY** of this tunnel was non-existent, yet this family of three had settled there. Brazil has already lost its ability to care for its children. How many countries are on the brink of the same tragic dilemma? How many of our cities present similar stories daily? The answer is too many.

To **CONGRESS, CLERGY** in **COMMUNITIES** of **FAITH,** and the **CATHOLIC CHURCH,** I say lead or get out of the way. Lead by forming a holy alliance to lead all of us to new visions and new responses. Pope John XXIII coined the phrase "the universal common good." Drastically limiting conceptions is for the common good of everyone. In Catholic teaching the common good defines what is necessary for everyone to thrive as a human being. We are so far removed from having the human community thrive today that we need a complete change of conscience to put in motion the adoption of new world wide comprehensive birth control policies immediately.

Even with solid factual data at our disposal it is hard to rate the Killer C's as to their level of negative impact on the environment. However, even Captain Cousteau is now dedicating his life to speaking about the **CONTINUING CONCEPTION CRISIS.** Over-population and over-consumption is the number one Earth family Killer. Therefore, it is necessary to offer controversial, radical or extreme thought for solving the crisis. **1.** Everyone could visualize voluntary human extinction as a motivation to change. (nonsense for most, yet a highly likely result of present policies) **2.** Governments allow no children to be born for ten years. (totally unrealistic, but probably the most humane and effective solution) **3.** Do not be afraid to speak truth to power about the folly of present policies regarding women's rights, birth control, abortion, family planning, and the allocation of funds. (realistic, but threatening and challenging) **4.** Use soap operas to teach about family planning. (being done successfully in many countries worldwide) **5.** Congress, Corporations and The Church could take the lead by funding programs that alleviate the **CAUSES** of **CONCEPTION.**   (possible and probable) **6.** Focus on stopping conception with clinics and accelerate the adoption of Earth's "other children" that need a nurturing loving parent or guardian. (realistic and necessary) **7.** Create small family clubs through the United Nations (being done successfully in the 4th largest nation in the world - Indonesia; should be mandatory worldwide and funded by the developed nations)

In other words, it is time to create radical changes for the sake of our children. Whatever it takes to bring about this conversion should be done, short of violence. A radical shift in our priorities, an environmental U-turn is the only thing that will give our children hope.

If we continue to allow 260,000 new mouths to ask for the world's resources, every day of every year, the human family will face challenges beyond our collective ability to handle. It's either caring **CONTRACEPTION CLINICS** immediately, or face the dire consequences tomorrow. Ninety-five million new consumers every year spells creations' eventual collapse.

CONSUMPTION

At the conclusion of the Earth Summit in Rio de Janeiro, the global environmental organization Greenpeace hung a huge sign saying <u>SOLD</u> across a two-hundred-foot globe. At this historic gathering the developed nations of the North were asked to carefully examine their consumption patterns, just as we were asking the less developed nations to examine their conception patterns. The "have nots" of the world believe we are robbing the Earth of its development capital by selling it off to the highest bidder. In essence, we of the North are caught up in what John Robbins calls the "cultural trance that is exploiting the resource base and teaching us to shop until the planet drops." To fuel our greed we rape the natural resources of the planet.

Undeniably we live in the greatest nation on Earth, but also the **CONSUMPTION CAPITAL** of the world. Alan Durning, in his book *How Much is Enough,* is well aware that Americans discarded over 32 million air conditioners, washers, dryers, stoves, water heaters and dishwashers in one year alone. He calls this and other patterns of our **CONSUMPTION CULTURE** leaving an "ecological wake that stretches globally." In the U.S. we toss away 57,500,000 soda bottles a day. Discarding something is a daily ritual for millions. We consume and discard, consume and discard to the point where almost everything is expendable. The ultimate symbol may be disposable diapers which we consume and throw away in the millions. We have opted for convenience over conscience and we consume without thinking of the consequences. For example: A family of four in California consumes approximately 700 gallons of water a day, while much of the world may spend hours procuring just a cup or two of this precious resource.

CALIFORNIA is the **CONSUMPTION CAPITOL** of the U.S. Yet, it is not enough to examine the lifestyles of the rich and the famous, though it is tempting to point the finger of righteous indignation at their display of limos, jet planes and opulent parties. The issue of excessive consumption is nothing new and not confined to Beverly Hills. "Like Ancient Israel," writes Dr. Walter Brueggeman, "we live in a world shaped by greed and consumerism of a royal urban elite." The acquisition and disposal of material possessions is an addiction for most Americans, so we need to turn the finger around and point it directly at ourselves. We are a part of the **CONSUMPTION CRISIS** because all of us consume more than we need.

Regardless of income or job, we currently use too much of the world's resources. Every purchase we make, every vacation we take, every mile we drive, and every letter we write has an environmental consequence. Our purchasing habits reflect our relationship to the natural world. A complete personal consumption critique is necessary by everyone. How, where, and how much of our money we spend is critical in understanding our personal role in the environmental crisis.

Henry David Thoreau, perhaps while sitting next to Walden Pond, reflected upon his life and commented: "A man is rich in proportion to the things he can afford to let alone." This is a good starting point for any **CONSUMPTION CONVERSION** process because nothing seems sacred anymore and we leave nothing alone.

We love buying stuff in our country and the success of the home shopping networks and the scores of holiday catalogues are the outward and visible signs of our inward addiction to consumption. But changing buying habits is not easy. We are pressured to the tune of 260 billion dollars spent yearly on advertising worldwide. We are taught to believe the engine of economic progress runs on the fuel of our spending and buying habits. Non-consumers are even called un-American by some. We have a deeply held belief in the endless supply of everything. Consumers are conditioned to believe that the elimination of natural resources will be offset by technology creating something else take their place. With these and other attitudes firmly in place we consume without conscience and the environment suffers. The truth is we are running out of natural resources, but our denial is so strong that we consume at ever increasing rates believing technology will provide.

However, all of us can adopt a better way of behaving as consumers. We can become experts on **CONSUMPTION CONVERSION.** Most of us are capable and eager shoppers already so we just have to learn the other half of the equation. This means switching the products we use, altering our patterns of shopping, consolidating and re-using resources, and living more simply. Changing to healthier and more earth-friendly brands, becoming eco-conservatives in our own home with water and electricity, and doing without some of the stuff does not have to be a slow and painful transition. Granted, tossing off old behavior patterns is not always easy, but with ready-made guides like *Earth Score* available to help us critique our consumption habits, we can aid the process. Unhinging ourselves from the consumption culture is actually personally satisfying. Ask a friend who has been owned by his or her possessions, and then finally breaks free from them. There is a certain freedom that occurs and perhaps a consumption anonymous group in every town in the United States could help millions kick the buying habit.

Personal consuming habits are also a question of justice, equality and morality. Why are you entitled to more of the resources than Wilson across town or Juanita across the globe? Is it because of birth, education, status, and income that Americans and others are able to consume to their hearts content? Many people believe that any or all of the above is justification for doing "what I damn well please with my hard-earned money." Individuals bristle when a suggestion is offered about the disposition of "their" wealth. I remember one evening looking at the plans of an 11,000 square-foot ski house in Sun Valley, Idaho which was going to be the couple's fourth home. They could not wait to furnish and decorate their new acquisition while I had to bite my tongue. This couple's consumptive lifestyle impacts the planet's resources far beyond the impact of an entire village in Africa. This may be the extreme of **CONSPICUOUS CONSUMPTION,** but many people covet that lifestyle; including middle American hard working citizens because achieving that style of living is seen as fulfilling the "American Dream".

Bill and Susan age 44 and 40 respectively, both work hard to maintain a home in the typical suburban town of Reseda, California. Like many American couples they are concerned about their quality of life. They hold two jobs so they can afford to raise their children with a particular lifestyle. They worry about the influx of immigrants into California and on the global scene they worry about the population growth in Asia. They are annoyed at the lax pollution standards that they have heard about in Russia and Eastern Europe. Daily they complain about the smog in LA and the graffiti and litter all over their city. What they do not think about is their own lifestyle and its impact on the environment. The truth is that their five children, or my three children will affect the state of the environment far more than a similar amount of children will anywhere else in the world.

Depending on the country of comparison, middle class children will consume from 3 to 500 times the natural resources of a child from another country during the course of their lifetime. Japan's citizens are the second most consumptive people in the world, but we still consume three times the amount of goods they do. Comparing the lifestyle of the majority of the world, especially in the developing nations, to one of Bill and Susan's children (and yours and mine) is like comparing the appetite of a lion to a mouse.

A middle class person from Anywhere, USA will drive 3-10 cars in a lifetime, discard tons of packaging from stereos, VCRs, CDs, etc. and use and throw away several appliances. They will read thousands of pages of material published on non-recycled bleached white paper, and toss plastic and clothing around in vast quantities that defy the imagination. One typical American child's impact on the planet is representative of our entire cultural emphasis on consumption. We are calmly robbing the future to pay for the so-called pleasures of the present and we are all complicit in this crime.

Yet, most of us continue to hold up the "American Dream" as one that is cluttered with possessions. We often even equate wealth and consumption patterns with the content of one's character. We have bumper stickers that say "the one with the most toys wins." Numerous examples can be found to indicate our enchantment with **CONSPICUOUS CONSUMPTION**. A few years ago the Los Angeles community marveled at the construction of a home that rivaled the Chateaux's of Kings. The owner was a person of humble beginnings who had the good fortune to create hit television shows. Rather than convert Bing Crosby's lovely turn-of-the-century home into a classy updated dwelling, the wrecking ball took over and leveled the property to the ground. Arising from the rubble was a 55,000 square foot mansion with landscaped gardens, imposing walls, rotating television cameras that sweep the grounds and a lot of empty rooms, because only two people live there. What was amusing for the press, a target of ridicule by others, and a draw for tourists became even more outrageous when the owner's wife's closet was discussed. The lady of the mansion, whose name coincidentally begins with C, had a **CLOTHES CLOSET** built to her specifications. This arena housing shoes, dresses, belts, purses, slacks and accessories is over 5000 square feet in size. One person's closet is more than double the size of an average home in our country.

I am sure this is the first time a person's closet has been discussed in a book about the environment, but closets may be a good place to assess our personal consumption patterns. Much of the world probably does not even have a word for closet. A large portion of the planet's population can transport what they own in a small cart, or even on top of their heads. Yet in our country real estate agents often stress the size, number and location of closets in the home as a strong selling point. Do we really need all that we have? Do our possessions own us? Can we convert to a simpler lifestyle? In the days of the ancient wisdom prophets held a mirror to the face of the people and asked them to examine their behavior. What is apparent today is that our personal lifestyle choices may predict the ultimate health of our nation, and eventually the environmental health of the globe. "We must go beyond dwelling on the protection of the environment," writes Friends of the Earth, "to address the underlying reality that the main obstacle is over-production and over-consumption in wealthy countries." Examining the simple daily rituals of personal habits often indicates our respect, or lack of respect, for the Earth. Reduce - reuse - recycle is the modern day trinity of the environmental community, but it must be attached to the themes of consumption conversion and curtailing consumption if it will really make a difference for the children.

Our cultural addiction to buying things and using up resources is fueled by the idol **COMMERCIALISM.** However, this affection has been met head on by a group that has declared September 24th as "Don't Buy Anything Day." A consumption fast would be an outward and visible sign that we recognize our consumption patterns are contributing to the decline of the environment. This day, for obvious reasons, will never be as popular as Christmas, but adherence to a day of not consuming creates solidarity with those around the world who have nothing in their closets and are struggling for some small degree of consumption parity.

In summary, **CONSUMPTION CONVERSION** is a vital first step. The environmental business expositions called Eco-Expo can give many of us an opportunity to see the next wave of environmentally sound products and companies in one place.   But the second and perhaps greater level of commitment to fostering a sustainable planet is to live more simply even if you have the cash to spend. Austerity and self-denial is not intended as punishment, but as a reward to those who will come after us. **CURTAILING CONSUMPTION** to a level of satisfying need rather than greed will be a significant healing C for the planet. There is great truth in the ethic of living more simply so that others may simply live. Weaving this into the developed nations environmental ethic will demonstrate to our brothers and sisters in the developing nations that we really mean business about a one world community of communities where the basic needs of all are met.

CHLORINE

"There is a chemical that is used to purify water in 98% of U.S. public water systems, to manufacture 85% of all medicines, and to produce 96% of all crop protection chemicals. That chemical is chlorine - used to meet the most vital needs of modern life," states Brad Lienhart, Managing Director of the Chlorine Chemistry Council. He goes on to make the argument that giving up this chemical would jeopardize the jobs of 45 million people and alter the way we live.

Is chlorine a magical wonder chemical? Let's look to Mosby's Medical and Nursing Dictionary: "Chlorine has a strong, distinctive odor, is irritating to the respiratory tract, and is poisonous if ingested or inhaled" Webster's Dictionary describes it as "a greenish-yellow, poisonous, gaseous chemical element with a disagreeable odor," yet we produce approximately 10 million pounds of chlorine each year. Chlorine is used to bleach paper white and to whiten our clothing, and it can be legally dumped into our water systems. Chlorine added to streams, rivers and our drinking water sources adds to the modern chemical soup and the results are often cancer causing dioxins. One chlorine manufacturer calls their product "White Magic" and sells it all over the world to unsuspecting consumers. It is not magic that is causing the most beautiful lake in the world to become polluted. Lake Atitlan in Guatemala is being killed with chlorine.

In addition, in the 1991 government study called <u>The Scientific Assessment of Ozone Depletion,</u> **CHLORINE** in the atmosphere was labeled as a major contributor to the break-down of the protective ozone layer. This affects every one of us. The Cancer Center in Houston reports: "If ozone decreases by 1 percent, UV-B radiation will increase by 2 percent, and the incidence of common skin cancers will increase by around 3 percent in the USA." Each year approximately 500,000 new cases of skin cancer are diagnosed in our country. But the rate of skin **CANCER** in **CHILDREN,** especially in Australia, is on the rise, and that is even more alarming. Because of the dramatic increase in the size of the ozone hole over the entire continent, Australian children must abide by the slip, slap, slop rule.   The warning to children: slip on a shirt, slap on a hat, and slop on sunscreen before you go outdoors and play. Skin cancer, however, is only one of the concerns of parents worldwide. An increased exposure to ultraviolet radiation causes cataracts and a weakening of our entire immune system. This is a heavy price to pay for white paper goods, swimming pools, and bleached white clothes.

But as consumers we can make informed choices. We can chose to adopt a consumption conversion process that leads us away from our chlorine chemical culture. Converting to using paper bleached by oxygen, or using non-bleached recycled papers, or paper made from corn, hemp, seaweed or kenaf is a way we can contribute to the health of the world's children, and help guarantee the health of our own children in this country. Today alternative papers and products are becoming more and more cost competitive. Products for the washing machine without chlorine bleach are on the shelves in our local supermarkets. Solar aquatic pool cleaners can reduce the use of chlorine chemicals by 80%. They even save you money by keeping the pool algae-free, and they are becoming more readily available.

Chlorine, as the gentleman extolling the virtues so clearly stated, still remains everywhere and in everything. Perhaps his own argument is reason enough to target chlorine for elimination; as we have done with Chlorofluorocarbons. Chlorine is so pervasive in our culture that its ultimate elimination might be a clue to unlocking many of the health problems we now face. Two of the major paper manufacturers have been shown the light by hard working environmental eco-conservatives and will be eliminating chlorine from their huge pulp mills. Like chopsticks, chlorine may not rank high on the hit list of environmentalists; however, the hidden dangers are often the most threatening in the long run.

CHLOROFLUOROCARBONS

In a 1994 issue of Worldwatch, Lester Brown writes about promising new trends in the environment. "Among the most promising developments in recent years are the wholesale dismantling of nuclear weapons, the decline in global military expenditures, the growth in bicycle production, the decline in cigarette smoking, and the dramatic reductions in **CHLOROFLUOROCARBONS (CFC's)**. This Killer C is a family of chemicals with long names that has been commonly found in our refrigerators, home and automobile air conditioners, fire extinguishers and aerosol cans.

Worldwide interest in CFCs resulted from the discovery of the hole in the ozone layer in 1985 by British Scientists stationed in Antarctica. Because of that discovery and the connection to CFCs the estimated global production has fallen dramatically. "The drop in CFC production," continues Lester Brown, "presages a decline in emissions.

This in turn will slow depletion of the ozone layer over a period of several decades and, if CFC use is phased out entirely, lead to eventual healing of the ozone layer." This is one of the few global environmental success stories. Conversion to alternative products is promoting healing, but we must remain steadfast because under the best of circumstances stratospheric ozone will not return to normal levels for many generations.   Here are four reasons why things are getting better. **1.** The Montreal Protocol challenged nations to ban the production of ozone depleting chemicals. **2.** Corporate conversion by companies like Hughes Aircraft to alternatives such as water soluble cleaners for circuit boards instead of ODC's (ozone depleting chemicals) challenged other companies to make similar changes. **3.** Environmental activists challenging companies that were slow in compliance or conversion procedures demonstrated that non-violent confrontation can raise awareness that leads to change. **4.** The key point in the success story is probably the fact that the general public had a tangible item, the aerosol can, that was known to affect the ozone layer. When manufacturers said they could make pump aerosol hair sprays without the harmful ODCs everyone applauded. Changing a simple everyday household item helped focus the general publics attention on the issue of ozone depletion.

Measuring the ozone layer illustrates how taking the Earth's vital signs on a regular basis is an important step in the restoration and healing process of our decaying environment. Thank Worldwatch for the yearly update on key environmental issues. Congratulations to the companies that are converting and phasing out harmful products. Challenge your colleagues in other companies to do the same. We will all breathe easier knowing the ozone layer damage is being minimized, and hopefully healed. **CFCs** are now officially removed from the Killer C's, even though their damaging effects will be felt for decades.

CHEMICALS

Headline Los Angeles Times July 10, 1991: "Pesticide Flows Into Shasta Lake. A toxic green river of pesticide poured into Lake Shasta early Wednesday as scientists carefully tracked the **CHEMICAL'S** potential threat to the state's drinking water supply and an army of frustrated officials from a dozen public agencies stood by helplessly, unable to stop the flow or **CLEAN** up the **CONTAMINATION.**" "Fourteen years ago, the future of 140 miles of

Montana's Clark Fork river looked as bleak as any science fiction writer could have imagined. Forming the largest polluted area in the nation some 50,000 acres of river corridor were littered with arsenic, copper, zinc, cadmium and lead. Heavy rains swept these toxic metals into the river. Toxic dust denuded a forest. Then came Superfund: Comprehensive Environmental Response Compensation Act, an ambitious plan to finance the cleanup of the nation's most hazardous abandoned dumps. $100,000,000 worth of cleanup is underway in and around Clark Fork. But for all that, surprisingly little has changed."

These are not isolated stories. Little has changed at Clark Fork, little has changed on our farms, little has changed in our waterways, and **CHEMICALS** are introduced faster than Wal Mart builds a new store. Estimates are that as many as 50,000 chemicals may be present in our air, water and soil; and in 1992, American companies alone admitted to the discharge of 3.8 billion pounds of toxic chemicals. The frothing chemical cauldron keeps being added to on a daily basis, and no one seems to know the full extent of the potential damage of this soup. Even the chemical industry has created its own 184-member Chemical Manufacturers Association with a "Responsible Care Initiative." This 10-point guiding principle outline has been in place since 1988. Created because of public concern over the proliferation of chemicals, the industry itself is scared to death about its own creations.

In a recent Natural Resources Defense Council report, co-released by CalPirg, The California Public Interest Research Group, there is indication that the chemical industry's belief and behavior does not always line up: "The powerful pesticide lobby has so far prevented the passage of a national law that will protect the environment and our health." In other words, chemicals are making us ill and killing us and the average citizen is not able to adequately protect themselves.

If the chemical industry wants us to believe that "Better Things for Better Living Through Chemicals" is possible today, they need to re-invent what it means to be a chemical company. Enthusiasm over new wonder chemicals has led to a corporate blind spot that is now beginning to affect the health of the planet's eco systems and their inhabitants.

E.B. White wrote: "I am pessimistic about the human race because it is too ingenious for its own good. Our approach to nature is to beat it into submission. We would stand a better chance of survival if we accommodated ourselves to this planet and viewed it appreciatively instead of skeptically and dictatorially." This is an indictment of everyone because most of us use some form of chemical daily, and many chemicals seem to beat everything they come into contact with into submission. Yet, as early as 1959 an eminent scientist writing in the Journal of Agricultural and Food Chemistry journal wrote: "We are going to have to do some very energetic research on other control measures, measures that will have to be biological, not chemical. Our aim should be to guide natural processes as cautiously as possible in the desired direction rather than to use brute force."

These quotations were referenced by another of our twentieth century Healing C persons, **Rachel Carson.** This pioneer, risk taking, "good science" person turned our attention to the potential damage from a chemically-based culture. Her book *Silent Spring*, written in 1962, might be re-titled today *"The Silent Century"* because we are now seeing the results of her predictions. We are approaching a time when the voices in many eco systems like coral reefs and wetlands are going silent. This silence often results from our continuing attempt to control nature chemically. A chemically-based culture is a concept conceived in ignorance and perpetuated in arrogance. The Earth's eco-systems can process our waste and many of our mistakes, but they have a hard time keeping up with the proliferation of chemicals that we carelessly pour into our life sustaining systems.

Nationwide we use over a billion pounds of pesticides a year and three quarters of this is used for agriculture. In California, the business of growing food is chemically dependent. We apply upwards of 300 million pounds of pesticides and herbicides a year; or one-quarter of this nation's total usage, much of which lands on the ground rather than on the crop. The argument postulated by the sellers of chemicals is that crop yield has been dramatically increased. Not only is that a fallacy when we look at cumulative long term results, (Quotation from Worldwatch Vital Signs 1993 "response of crops to additional use of fertilizer is diminishing") but the death of the soil and illness of the workers is a by-product. I have been driving between Los Angeles and San Francisco for 20 years along Route 5 and through the middle of our rich, fertile agricultural lands. In the good old days the windshield was covered with bugs at the end of the trip and I would complain out of ignorance. Today I travel the 350 miles and never wipe the windshield once and complain out of knowledge. Bugs represent life in the soil and create the necessary nutrients for growing our food, but today they are disappearing because of the over-use of pesticides and herbicides.

In addition, chemicals directed at the pests on our food often end up in our water. In 1992 an additional 1500 wells were closed in California because of contamination from agricultural runoff which is not regulated by any agency anywhere. But when we point the finger at others, we have to aim it at ourselves also because we put chemicals down the drain daily. Household detergents, auto cleaning products, and a variety of common substances all contain chemicals that can contaminate our water supplies. In a TV special called <u>Race to Save the Planet,</u> the commentator was speaking about the quality of water in Europe. "We all drink contaminated water" was her matter-of-fact summary statement. Closer to home, cities in Iowa give the nitrate (fertilizer) count in the water on the local evening news, and since 1987, 10 herbicides and 4 insecticides have been routinely detected in Iowa rainwater. In the NRDC report, referred to earlier, over 2,631 violations of the safe drinking water act were reported in California in one year. These violations alone affected over 16 million people. Americans take for granted that our water is clean and pure for drinking and swimming, but that is no longer the case. Even coastal waters have been **CONTAMINATED** by **CHEMICALS**. In Humbolt County, California a few years ago another major paper manufacturing corporation was cited with 40,000 violations of the clean water act. Young men and women were getting sick surfing in brown water. A major law suit was filed and won by the Surfrider Foundation. In communities all across the country storm drains deposit polluted runoff daily in bays and rivers. In the huge Santa Monica Bay, that is the water playground for millions of Southern Californians, lifeguards now shower after every swim. After a storm parts of the bay may be closed for weeks. Water, water everywhere, but we can not drink it, we cannot swim in it and in parts of Europe the water is so contaminated you can not even cool machinery with it.

This sad condition of our water supply prompted the United States Public Health Service to test our internal environments for contamination. They tested for 100 toxic chemicals. Five people had all 100, including DDT, and 50% had at least 35 different chemicals. The scary part of this relates to our early education in basic chemistry. As students we were warned never to mix chemicals because the results might be catastrophic. Yet many of our children already have the ingredients of an unknown internal **CHEMICAL COMBINATION** that represents potentially harmful long term consequences. The ingredients of this mixture come primarily from our food, water and air. The three ingredients essential to a creature's existence are now a threat to the same creatures that depend on them for life itself.

It is estimated that by the age of 5 a child will have consumed 7.5 pounds of chemical food additives. The NRDC has found that pesticides are present in at least 35% of all analyzed food samples. Our children are also encouraged to play outside on our green lawns and fields fertilized with a variety of pesticides. Also, there are 25 million adults who play golf on the 3000 square miles of fertilized golf courses in the United States. Frankly, more chemicals go on our manicured adult playgrounds than on our food, but both are excessive. We also create sport and game time at our schools, both in suburban communities and inner city sites. All this running and exercising, that is supposedly good for our development, is often done in places where our air is filled with lethal chemicals. Our children's lungs become the filters for the pollutants in our atmosphere. In other parts of the world the situation is worse. Some families in Mexico City will not allow their children to go to school there anymore because of the pollution. If you can believe this; they come to America because of our "'cleaner air."

In addition, many world class athletes can no longer train in their contaminated home-towns. Athletes should form coalitions with clean air advocates because their ability to perform at peak levels depends upon them having healthy environments.

A sensible idea might be that in addition to giving our children a primer on the history of the industrial revolution, or our athletes weight training guides, we might serve them well by offering a subscription to Safe Food News or The Clean Air Campaign literature. Personal knowledge about alternatives may be the first line of defense in our **CHILDREN'S COUNTERATTACK** on **CHEMICALS.**

Another line of defense is **CONSUMER** action designed to encourage people to buy non-toxic products, drive their cars less, and strive for a chemically-free home environment. Also, teaching our children how to grow their own food in cooperative community organic gardens may give them the skills necessary for the restoration of their own health Teaching them the value of composting may revolutionize agriculture. Composting and gardening should be taught beginning in the first grade in school Quoting from the Walt Disney Company's environmental activities report: "Walt Disney World has created a state of the art composting system to handle its sewage sludge. This successful program has recently expanded to accommodate landscape and food waste as well. The compost currently produced is being used as a soil amendment by the Horticulture Department." The vast Disney empire sees the worth of composting. I can see Mickey and Minnie teaching our children how to become chemically free and composting competent. Habits can be changed, and with a little help from our friends like Disney and Earth Save, an environmental organization promoting a chemically-free vegetarian diet, our entire educational food delivery system could have safer, healthier foods for our children. With clean air and clean food, clean healthy children can become commonplace again.

CHEMICAL CLEANSING may be a slow and heart wrenching process that will be blocked by farmers, oil, fertilizer and seed companies. But the conversion principle still applies. Start with yourself, challenge a friend, talk to the greens keepers at the golf course, buy organic fertilizers for home, convert an office to non toxic products and then tomorrow the consensus of the world will be that we only really need a few of the thousands of chemicals on the market today. My dream is that chemical companies will see the wisdom in the conversion to a less chemically dependent and more organic culture, in part because their own children are not immune to the illnesses so prevalent today.

CANCER

While the cancer industry ignores the cries of "fix the environment, **CURE CANCER**," the incidence of cancer in America continues to rise by 2%. We have spent 8 billion dollars trying to find the cure, but none is in sight. In this year Americans will suffer with the reality of the death of over one half million people from cancer. Cancer concern is at the heart of everyone's family health agenda, yet too little is done to connect the amount of cancer with environmental contamination.

Why not offer a large percentage of the millions of dollars presently funding research to environmental organizations who are trying to gain compliance with clean air and clean water legislation? In reality, front line eco-conservative activists may be doing more for the health of the population than all the hospital research combined because our basic elements that are supposed to sustain life now contain chemicals that cause cancer.

Concern about cancer dominates the thinking of many segments of society, especially within the "**CHEMICAL**

CORRIDORS " in our cities or "chemical valleys" in the rural parts of our nation. Cancer scares us to death, but too often we abdicate responsibility to the cancer industry to find the cure and condemn the activists who challenge the industries that are polluting with cancer-causing substances. Every citizen must put on the badge of courage and become a **CANCER COP**. Learn about the environment from a selfish, self-centered point of view, because your life and your children's lives depend on it.

Each one of us can take an interest in our local environment, especially as it relates to our children's health. Two years ago the winner of the top prize of L.A. Magazine's Environmental Pride Awards was a grandmother named Stormy Williams. She has fought for years to protect her rural community from a wide variety of un-healthful factory emissions. She has been threatened, ridiculed, and finally honored for her courage. "I've made a difference," she says proudly, "but not a lot of friends in the industry." Several years ago the Concerned Citizens of South Central L.A. said no to the construction of the Lancer Disposal Site in their community because of health concerns. Granted, the siting of toxic producing factories or industries is a difficult task, yet the air we breathe must remain clean or we will all have environmentally-caused health problems one day. People everywhere, especially in communities of color, are talking and doing something about environmental justice and toxic racism. "Not in my backyard" should mean not in anyone's backyard, unless the industry does not adversely affect the health and well- being of the neighbors. Today there are over 5000 non-profit environmental groups voicing their concern across the United States. Many of them are fighting for the health of the children, especially in the inner city. They should be applauded, rewarded, and financially supported by eco-conservatives everywhere. Community activism must be balanced with personal changes. There are Healing C's that affect our own internal environments. Our personal food choices have a direct correlation to our health and well being. Carolyn Clifford, Chief of the Diet and Cancer Branch of the National Cancer Institute, emphasizes that many scientific studies have demonstrated that eating greater amounts of vegetables, fruits and grains may lower the risk of getting cancer. Several of these food classifications begin with the letter C. Nutritionists and doctors are encouraging their patients to eat more **CRUCIFEROUS** vegetables: broccoli, brussel sprouts and cauliflower. We are also advised to eat lemons, limes, grapefruits and oranges: members of the **CITRUS CLAN.** The third food group is called **CAROTENOIDS**: orange and yellow vegetables. Carrots, sweet potatoes, pumpkins and papayas, as well as fruits contain healthy quantities of betacarotene. No one claims that this is a bullet- proof formula for a cancer-free life, but a diet based upon these food groups may help our internal environment; especially when these vegetables are grown with *Seeds of Change* organic seeds in composted organic soil with no chemical additives.

We are not helpless in the face of cancer's growth and the environment's decay. The late **Norman Cousins,** another of our Healing C persons of the 20th Century, encouraged people with cancer to use humor and a positive mental outlook to help in the healing process. He was well aware that a good external attitude helped maintain a good internal environment. He knew that stress and anger, on the other hand, may potentially harm the healing process. In similar fashion, a positive attitude about helping restore, heal and preserve the external environment is essential today. Granted, feeling anger at the culprits of environmental decay is quite normal, but usually counterproductive. Instead **CHALLENGING, CONFRONTING,** and **CREATIVELY** finding common ground with our "enemies" can produce change. Even just reading about the Killer C's in this book can lead to anger and depression, but taking care of our internal environments strengthens us to make a difference in the world. A positive outlook on life is an ally in combating cancer and restoring a sick planet at the same time.

Members of the healing professions are also a very important component of an emerging environmental awareness in this country. A significant coalition that will help deter the spread of cancer and fix the environment is an alliance between the helping professions and environmental activists. This team would accelerate increased awareness about the connection between cancer and the causes. Doctors, healers, psychiatrists, psychologists and counselors, by weaving a discussion of the environment into their office visits, will help heal millions of sick people. Sensitivity to diet, attitude and lifestyle coupled with keeping air, food and water clean will become the leading health delivery system in the next few years. **"CANCER CLUSTERS,"** or unusual amounts of cancer in one area, happen for a variety of reasons. A coalition of environmental cancer cops and healers might be just the antidote researchers have been trying to find for years. "Fix the environment and **CURE CANCER**" should be stenciled on every automobile, water delivery system, smokestack, and factory in the nation. "Fix the environment and **CURE CANCER**" posted on the front door of every doctor's office and health food cooperative would remind us that each of us can take responsibility for our own health and long-term well-being. But, this is only part of the battle for millions.

CIGARETTES

Three hundred children under the age of 18 start smoking **CIGARETTES** in **CALIFORNIA** every day. This means that their developing internal eco-systems do not stand a chance of long term good health. Perhaps even more frightening is that an adult's right to smoke is compromising the health of our nation's children. The external physical environment is no longer benign for our kids. Any sane person would not put their child's face in front of a tailpipe of a bus taking their child to school, yet there are rational adults blowing smoke in their children's lungs in homes, cars and restaurants across this land every day of the year. Many caring parents are adamant about their right to blow smoke wherever they choose, even at the expense of their beloved offspring.

This is an environmental health issue of crisis proportions and **CITY CHILDREN** have a triple whammy to face. When we breathe clean air our lungs expand and we remain healthy. When the air is polluted our physical well-being is at risk. Polluted air and smog fills our cities' corridors because of bumper-to-bumper cars, diesel trucks, busses and factories. Our children also have to contend with **CIGARETTE COMPANIES** advertising a fellow named Joe Camel who shows how "cool" it is to smoke. If that is not enough, second-hand smoke is adding to the dilemma facing our children. Children have no choice and their growing lungs are contaminated by the decisions adults make regarding the use of tobacco. This is reason enough to bring the tobacco industry to its knees. The irony is that our adolescent education always emphasizes taking care of our bodies. Health teachers do not encourage smoking because they know the consequences. Little League coaches and American Youth Soccer coaches teach the same message as major league managers: treat your body well and do not smoke anything because the health of the internal environment contributes in a large measure to success on the playing field. We only get one body and if we pollute it, it will not last long. Yet, millions of American adults provide terrible role models for the children as they keep on smoking.

The tragedy is that those who champion the use of cigarettes encourage internal environmental suicide. Advertising smoking anywhere promotes self-inflicted death. We punish Dr. Jack Kervorkian for medical-assisted suicide, yet we condone tobacco lobbies and elect politicians who block strong anti-smoking legislation. We cry for the death penalty of murderers in our country while we pay our cancer company executives millions of dollars a year. Governments and citizens are forced to spend precious dwindling resources for anti-smoking campaigns while tobacco companies hire top creative advertising firms to convince people to purchase cigarettes. What a crazy message to send our children. As a conservative I am appalled at this waste of resources, double standard, and erosion of our children's right to a fair and sustainable future. This **CIGARETTE CONSPIRACY** is an environmental killer of drastic consequences and must be dismantled step by step.

We know that cancer has many relatives, but only one first cousin: cigarettes. Cancer's ally is a friend in the form of four inches of tobacco rolled into a cigar or cigarette that in 1981 was smoked a record 650 billion times in this country alone. Today the estimates are that 500 billion cancer sticks are consumed by 40 million addicted and denying people every year.

The tobacco industry knows the truth because they are taking their poison elsewhere. Forced to label their product here, and with an increasingly aware population in America, the **CANCER COMPANIES** are knocking on new doors around the world. Uneducated about the dangers of tobacco, unsuspecting global citizens will now be targeted for slick Madison Avenue advertising campaigns. The "moral" CEOs are lining up on the borders of China, Eastern Europe and Russia. The R.J. Reynolds Tobacco Company is building new plants in Russia to produce 22 billion cigarettes a year. The U.S. market has dropped 15%, but we have increased production 5% to meet the demand from overseas. We are exporting another environmental cocoon as fast as we can that will wrap other parents' children in an accelerated death pattern; and we label this free trade and good business!

In the Tobacco Industry hearings before Congress the executives preached that cigarettes and tobacco smoke was safe. If they really believe their own propaganda then tobacco industry executives should be willing to smoke three packs a day of their own product while giving up their health insurance. They should also encourage their own children to smoke a pack a day of their product beginning at the age of 12. If tobacco was this safe to smoke then Congressman Waxman's hearings on the tobacco industry would no longer be necessary. But we know differently, and so do the tobacco companies.

The public would be better served if there is **CONVERSION** to **CASH CROPS** that are good for the environment. Products such as kenaf, Paulownia trees, and a variety of other fast-growing, high-yield, non-pesticide-intensive agricultural products can replace tobacco growing and provide jobs. Congress can offer financial incentives for immediate compliance with conversion plans and reduce health costs.

The long term **COSTS** of a **CULTURE** addicted to cigarettes is beginning to be understood by the health care profession, the insurance industry, the environmental community, and even the athletic world. It makes little sense trying to sustain eco-systems in the natural world when individuals fail to understand that they too are fragile interconnected systems that need good health care management. It is an oxy-moron to describe a smoker as an environmentalist. He or she may care about the external environment, but failing to preserve the system that sustains one's own life makes no sense. If our bodies are the household of the Creator, as many scriptures emphasize, conservatives must join in the coalition to ban cigarettes so that the individual temples of the Creator are not destroyed. In summary, the **CHILDREN** of the world who are enveloped in the growing **CIGARETTE CONSUMPTION CRISIS** need our strong voice because their, not our, long-term health is in jeopardy. Voices in power need to hear the cry of the children about cigarettes and help them avoid them forever, especially the Congress persons in tobacco growing states.

CONGRESS

The **CONGRESS** of the United States of America holds the key to conserving the precious resources of the United States. **CAPITOL HILL** and the **CONGRESSIONAL LEADERSHIP** can either strengthen or weaken existing laws that protect our threatened environment. Congress certainly controls the attention of environmentalists who are concerned about the Clean Water Act, the Endangered Species Act, the Rivers Bill, and the Clean Air Act. The 104th Congress may be the most scrutinized body of elected officials ever because this Congressional delegation will hold the children hostage if they fail to protect and preserve the legislation that guarantees them a clean and viable future.

Eco-conservatives in Congress, those elected and charged with preserving our great land, have the option to lead or to get out of the way. Issues related to water, air, soil, tobacco and inner city environments should not become partisan political footballs, but the heart of the Congressional ethic embraced by both parties.

A good place to start is with our water. America depends upon clean and safe drinking water, yet wells are polluted, aquifers are drying up and water delivery systems are falling apart. The Safe Drinking Water Act of 1974 was created to protect us. There are 200,000 public water systems in the United States and keeping them all safe is a challenge. **CONGRESS** has the option to guarantee across the board **COMPLIANCE** with EPA standards. Or, they can ignore having an environmental ethic and undermine the strong standards "on the books."

Congress will author roughly 15 new environmental laws each session while reviewing laws, treaties, and declarations already in place. One such treaty is the Amsterdam Declaration that we signed in 1989. It sets a goal of providing contraception to every couple on the Earth by the year 2000. The U.S. share, in dollars, is $650 million a year, or the equivalent of 25 jet fighters. We signed the treaty, along with 79 other nations, but have yet to commit one penny to its implementation.

Another example of a law in place relates to the Amendments to the Clean Air Act written in 1990 to regulate the 12 million marine engines that are responsible for 700,000 tons of pollutants to the air every year. Congress persons, like everyone else, love to go boating, especially with their children, but marine engines pollute both the water and the air. We hear very little about this increasing threat to our water systems and the health of our children.

It is prudent today to Adopt A Congress person. We can also create a **CHILDREN** to **CONGRESS** letter writing campaign to ensure that the laws of the land, and the treaties of the world, are enforced or adhered to on a comprehensive basis. Congress is a special body of people who deserve our undivided attention. Congress has been entrusted with the responsibility for the management of the American household and how well they do their job affects the health and well being of our children.

Also, Congress can no longer hold on to the myth that environmental regulation means loss of jobs and a poor economy. "The great majority of economy-wide studies show a small positive effect of environmental regulation on overall employment," states the

Economic Policy Institute's recent study, <u>Jobs and the Environment: The Myth of a National Trade-Off.</u> Clean technologies will create new jobs in large numbers as we move into the 21st century. This conversion should be promoted by every member of congress.

Issues of violence, jobs, and health care all relate to a transition in our society to a safer environment. Americans are **CALLING** for a **CONVERSION** to a sustainable economy where the common good is served. People are tired of Congress serving Congress and special interest groups. It is my belief that conversion to an environmental ethic will provide the context for transition to peaceful technologies, environmentally correct jobs, safer cities and streets, and healthier people who take responsibility for their own internal environments. The **CONVERSION** of **CONGRESS** to serving the common good is as important as the conversion of corporate leaders and clergy from Connecticut to California. But, we have to let them know we care about the health of our children, eco systems, and all in creation. A letter, or a long distance phone call to our adopted congress person gets results. Our calls of concern, especially about the legislative agenda, may break what Katy McGinty, the White House environmental advisor, calls the "cynicism within Washington about the environment." **Call Congress Collect** and remind them of their "contract with the children."

COMMUNISM

Communism, or the communist party in the former Soviet Union and satellite nations, brought on an environmental nightmare, the consequences of which the human family, the animal kingdom, and the natural eco systems are woefully unprepared to handle. Environmental eco-cide has happened in many parts of Russia. However, the closed system called Earth does not have an escape hatch over the Soviet Union, and their legacy of environmental neglect impacts all of our lives. Since the 1930s, communism's waste has been thrown into the atmosphere, tossed onto the land, and discarded into the water. One lake has even been labeled the Atomic Lake because the water is so **CONTAMINATED** with strontium nuclear waste that the danger will be around for thousands of years.

Ecological concerns were pushed aside with the broad sweep of the communist party's ruling hand in favor of what was affectionately known as "gross output." Communism's success was measured by how many tractors, factories and machines were produced. Ironically, at the same time there was a gross output of toxins, effluents and pollution which signaled wholesale disregard for air, water and land and the creatures that depended upon them being clean for their health.

One comic-tragic example is how the former communist leadership treated their diesel trucks. Diesel fuel has always been in great supply in the Soviet Union. The problem is that the trucks are hard to start in cold weather. The communists solved this throughout the Siberian region by leaving them running 24 hours a day, 365 days a year! Thus, they never had to start their trucks. In addition, hundreds of factories with old technologies continue to operate and are inefficient users of energy and gross polluters. The closer we look the worse it is.

The once majestic Aral Sea's waters have been drained to irrigate desert land to grow **COTTON** with huge quantities of pesticides. The Soviet Union has practically ruined the fishing industry as the sea recedes upwards of 40 miles a year. Communism's legacy of disastrous decisions affecting their own environment stretches to Siberia and Mongolia. Lake Baikal, the source of 80% of the nation's fresh water, and home to unique plants, fish and a fresh water seal that exists nowhere else in the world is under attack. Belching smokestacks cover the fir trees in a blanket of death. Pulp factories discharge their waste into the crystalline blue waters of this once pristine lake. What is sad for all of us is that another significant piece of creation's puzzle is disappearing. The greatest threat of all may be the radioactive waste moving through the rivers of Outer Mongolia towards the Arctic Ocean. Stay tuned on this one because it signals perhaps the greatest environmental threat ever. Toxic water knows no geographical boundaries because all the water systems of the world are connected to form one big pond and a pond is a closed system. Polluted water from the farthest reaches of the globe eventually mixes with the water in our local harbors. Therefore, concern mounts daily about this tragic communist legacy.

Where is the magical cure for a culture of communism that disregarded the wealth of the nations' natural resources for decades? The Healing C for communism may simply be a **COMMON CONCERN** or common ground of the two most powerful nations on Earth. The U.S. and the combined states of Russia have a common enemy called **CLIMATE CHANGE.** The global healing C may be the environmental **COOPERATION** of our 2 **COUNTRIES.** Technology transference, joint conferences on environmental remediation programs, and a clear path of change will mark the new millennium between our nations and hopefully the nations of the world. The cold war must now become the **COOPERATIVE CHALLENGE** to meet head on the drastic environmental changes occurring throughout the globe.

Cleaning up emissions from our polluting industries and sharing our technologies with Russia is another cog in the environmental wheel of progress slowly beginning to move. Seeing the communists as people with a common purpose and common ground will hasten the conversion process. A contaminated Russia will eventually contaminate the world. All we have to do is remember **CHERNOBYL**.

Thankfully, many of the young **COMRADES** of Russia no longer tolerate policies that create environmental disasters. Environmental protests are occurring throughout the republic and have led to conversion almost overnight. The people have awakened and now changes are happening and the policies of the past are disappearing. Now a cross-fertilization of ideas and strategies among and between cooperating environmental organizations across the globe can help restore the entire planet. There is one consequence of the Communist party's decisions that can not be met by good will, technology or conversion of the heart. There exist contaminated manufacturing towns where people were treated like animals. Because of chemical pollution and lax standards about nuclear waste, women are giving birth to grossly damaged babies. Ask the former communists of Semipalatink about the environmental hell created in their community. The entire environment of the region around that industrial complex is a death trap because the land is contaminated forever.

CHERNOBYL was a reminder to the Russian people that even hard work and new technology can not remedy some catastrophes. Communism's healing its own environment may be wrapped up in bringing to the attention of the world community the folly of a nuclear policy. Unless we learn to dispose of the waste generated the environmental consequences are too great to continue relying on nuclear power as an energy source anywhere on the planet.

Perhaps there is a silver lining in this Killer C that we have not acknowledged. There has always been the belief that nature has the capacity to heal itself. The nuclear disaster in Chernobyl coupled with the manner in which they have disposed of their toxic waste in the former Soviet Union dramatically indicates how Mother Nature needs our help in restoring the eco systems. In addition **COMMUNISM** may have inadvertently hastened the world wide conversion to sustainable energy programs based upon harnessing the sun and the wind because of their horrible experiences with the alternative. Their gift to us may be the lessons we have learned without having had to suffer the consequences.

Our response as a people must be compassion and not condemnation. We must reach out to the people of the former Soviet Union in a special way. The Russian people, the people of Romania, and the people where communism has all but died, need to know that the healing of their environment is as important to us as it is to them. It is in our own self-interest to know that there is no more radioactive contamination from Russia. It is for the future health of our children, as well as theirs, that we transfer technology that can help clean up their environment. Jobs can be created by the thousands when we think of communist countries as sources for our environmental companies. Conversion of our hearts to accepting their plight as part of our responsibility will begin the process toward environmental peace in the world. Today we do not have that peace, but tomorrow we must. It can begin with environmental dialogues across the oceans.

CONFERENCES COMMITTEES CONVENTIONS

We have to be cautious about these Triple Cs because they gobble up our time in great segments and too often simply recycle ideas, shuffle papers, re-work reports, create non-binding proposals by the bleached tons, and contribute to the contemporary chatter-chatter-chatter syndrome. If Nero fiddled while Rome burned, many environmentalists and others will attend conferences while the planet implodes and feel good about doing so. Some conferences are vital, others are busy work and many contribute to the problem rather than finding the solutions.

The caveat is simple. If you are going to attend a convention, conference or committee meeting think about the consequences of your time invested. Secondly, if you are addicted to the Triple Cs because they are woven into your way of doing business today, at least design every convention, committee, or conference as an opportunity to bring people together with two main purposes: 1. Create a way of empowering people to build strong coalitions that affect change 2. Build in opportunities and demonstration projects to foster the conversion to a personal and institutional environmental ethic.

Every time two or more people are gathered together **COMMON GROUND** and **COMMON VALUES** must be found. The issue of building towards a personal, institutional, and ultimately a worldwide environmental ethic should be at the heart of every gathering, regardless of the subject matter or prime focus of the conference, committee, or convention.

That means that the choice of paper, food, site, activities, accommodations, entertainment and agenda should reflect an environmental ethic. If the AMA is meeting in Hawaii, or the United Auto Workers are meeting in Las Vegas, or if the staff of the National Audubon Society is meeting at their state of the art environmental headquarters in New York, attention to the linkage of the environment and the day's business must be constantly in the forefront of everyone's agenda. We must always ask in the context of every meeting: How are our decisions and actions affecting the health of the environment?

For me personally there has been an interesting alternative to the Triple Cs. Seeking connection to activists who are at the grass roots changing and challenging policies that are harmful to the environment is a worthwhile enterprise. A valuable action might be to call an environmental, business, church, or community leader who believes in the value of an environmental ethic and ask to sit at their feet for a weekend, a day, or an hour.

One of the best hours I spent last year was listening to Ed Begley Jr., speak about "walking your talk" at the Earth Service environmental roundtable. One of the most interesting 3 days of my life was when I went to Washington, D.C. and lived with Mitch Snyder, former advocate for the homeless, in his Center For Creative Non-Violence. One of the most enlightening 6 days of my life was when I went to Riverdale on the Hudson to sit at the feet of Father Thomas Berry, author of *Dream of the Earth*. One of the most educational 21 days of my life was when I studied about the natural world of Costa Rica with two British scientists on an EarthWatch project. One-on-one connections can replace countless hours of conferences, committee meetings and conventions. Simply observing or following a person who believes in and lives the environmental ethic will accelerate personal conversion rates dramatically. The healing C is **CONNECTION**. Connecting to others who are passionate about their life and work rubs off on even the most hardened skeptic. Communion with creation balanced with connection to people whose beliefs and behavior line up is a life-long gift. Take an activist to lunch, and listen! Once the connection is made to creation and people deeply committed to their work, one can guarantee that you will never be the same again.

CHURCH

Listening to nature is more important than ever because the ecological balance continues to be upset in the Garden of Eden, a metaphor for the Earth. The religious community must listen to what the Earth is saying and communities of faith must recruit new leaders to take us to the promised land called a sustainable creation. The clergy must lead or get out of the way, because the people are rising up and asking for direction, especially spiritual direction related to a creation-centered theology. Justice and equity for all in creation is the bottom line. California State Senator Tom Hayden **CHALLENGES** the **CHURCH**; "Where is our moral covenant with nature, tied to our covenant with God? We need the voice of religion on these issues."

Fr. Thomas Berry warns that "the real tragedy is that the religious and spiritual persons themselves remain unaware of their need to provide for themselves and for society a more significant evaluation of this larger **CONTEXT** of our lives." Yet the clergy, more than any other segment of society, know the value of a healthy creation. Clergy are the first to go on retreats to quiet places set in nature. Clergy extol the virtue of sabbaticals, or their vacation time to renew and refresh themselves. Contemplative prayer, meditations and rituals of healing are often connected to natural places. The context of personal renewal is often nature, but clergy voices are not heard when it comes to preserving those rivers, lakes or mountain retreat areas so vital to their ministry.   Sermons on eco-systems and species preservation are as rare as the black rhino in Africa. Voices of concern like Dean Morton of St. John the Divine in New York, who admits that "only in the last few years has the church acknowledged the depth of the ecological crisis and agreed that it must be at the center of the church's life" are few and far between.

But a church without a viable creation is like a human being without a soul. The loss of clean water means the rites of purification or baptism become invalid. Polluted air means we have contaminated the breath of God and that breath gives everything life. All is connected, and yet everything is coming apart at the seams and the pieces of the Earth puzzle are disappearing as fast as a ticking clock.

Creation and its creatures need the healing C of **CARING CLERGY** worldwide who will become guarantors of their own biological region or section of The Garden. It is not enough to simply talk about **COMMUNION** with **CREATION**. We must, like our Native American teachers, become genus loci, or guardian spirits of everything God has entrusted to us. The church, long before the environmental community emerged, was asked to guard air, water, trees and animals. In Genesis the human family was collectively asked to till and care for the Earth. All leaders of faith communities are asked to keep and serve life in abundance. This means the ingredients which comprise a healthy garden need to be guarded and preserved. Therefore, the first job of everyone is to be good gardeners and clergy are to be the best gardeners. Today, it seems most clergy have been replaced as guardians of the garden by the Rainforest Action Network, or the Clean Air Coalitions, or the Endangered Species Coalition, or the World Wildlife Fund. It seems they are the ones reading the wisdom of the ancient teachers and doing something concrete to protect the garden. Clergy must listen to these modern day prophets. Frankly, **CARING** for **CREATION** is a lot more important than guarding rules, regulations and church policy. None of these ensure life in abundance the way clean inner cities, healthy water systems, contamination free farm lands, intact rainforests, and the diversity of species do.

Thus, clergy and environmentalists are the partnership for the new millennium. They share more common ground than anyone on the world scene today. This coalition needs to come together as the strongest, most effective body of human beings ever formed. Their collective voice will help strengthen the resolve of everyone else to preserve and conserve the elements of creation._ "The next thirty years will be the most serious ecological crisis in human history," warns Dr. Peter Raven, biologist and director of the Missouri Botanical Gardens. This is why science, religion and environmental activists must raise up leaders overnight to lead us to the promised land of a healthy Garden. Seminary's nurture, guide and groom their students in order to be able to take leadership roles in the community. As leaders clergy must understand that the theology of the future will connect all scriptures to the Earth and its systems, for they are part of the whole human story.

In addition, every community of faith (church) needs an eco-conservative activist on the board of the local congregation. A call should go out in every community, beginning in the U.S., for a

CREATION CONSERVATIONIST to join the synagogue, church, mosque or meeting. Their role would be educator and mirror bearer; i.e., the one who holds the institution accountable for its actions. At the same time, every active environmental organization should advertise for a clerical representative on their board of directors to bring a spiritual voice connecting them to the environmental ethic.

If this mutually enhancing partnership became an integral part of church life then the words of Basil the Great would begin to resonate in the houses of worship throughout the world: "I want creation to penetrate you with so much admiration that everywhere, wherever you may be, the least plant may bring you to the clear remembrance of the Creator." Or, perhaps the words of the Native American Chief Standing Bear of the Ogala Sioux would find an appropriate place in the worship of our modern church communities. He wrote: "The Old Lakota was wise. He knew that a human's heart away from nature becomes hard. He knew that a lack of respect for growing living things soon led to a lack of respect for humans too." Inclusive attitudes like this enhance traditional scripture and broaden our way of serving the Creator.

Probably the most important influence on clergy is their congregation. 52 weekends a year, communities of faith gather to hear the spoken and written word of God. Clergy usually select their own sermon topics because they are the designated, though not the only, spokespersons for God. The sophisticated laity, however, expect guidance in trying to understand the difficult issues facing all of us today and a dialogue between leader and listener is vital.

First of all, clergy must attempt to guide their **CAPTIVE CONGREGATIONS** away from an anthropocentric (human or man centered) world view. Humans are not the center of the cosmic story; we are merely parts of the complex web. Russell Train, a committed church person and the first director of the EPA says that "the remaking of the environment must begin in small yet significant ways by individuals who take seriously the ideas that the Earth is God's and not theirs to do with as they please." Yet too often the preaching and teaching from our communities of faith relegates the book of Nature, the Garden of Eden, and God's other creatures to minor league status. Clergy must reference all of creation as part of God's Kingdom. Everything God created is sacred, not just what we choose to sanctify.

Secondly, clergy are natural born eco-conservatives and disciples. They are called to conserve the values, the common ground of all peoples, and the gifts of God that sustain life. But, they need to remain as learners and listen to the wise counsel of others in order to be able to lead the children to a sustainable land. The promised land for me is an Earth that can be sustained forever so that generations unborn may enjoy and continue to preserve the fruits of creation.

The optimum situation for our communities of faith is when the **CONCERNED CONGREGANTS CHALLENGE** the many **COMPLACENT CLERGY** and the clergy respond by creating programs that foster an environmental ethic in the parish. In 1993 a document called <u>God's Earth Our Home</u> was sent to 55,000 congregations of faith. This well-written, informative, and comprehensive document is capable of educating and guiding congregations to a strong environmental ethic. Sadly, too often documents like this sit in the stacked paper file on clergy desks. Laity need to ask their clergy about this wonderful tool for congregations. An appropriate action for the National **COUNCIL** of **CHURCHES,** which produced the document, would be to partner with Green Corps canvassers. They could go door to door to every church, mosque and synagogue in the United States and encourage compliance with the program of <u>God's Earth Our Home</u>.

Another thought relates to resolutions passed at communities of faith conventions. These thoughtful, but often non binding edicts could be printed and hung on the front door of every church or synagogue of the corresponding community of faith in America. For example, the 1994 General Convention of the Episcopal Church Resolved: That the Executive Council through its Social Responsibility in Investments Committee be directed to screen its investment portfolios for environmentally responsible corporate behavior inside and outside the United States, and to pursue corporate dialogue and shareholder resolutions with those companies wherever they are located to assure compliance with environmentally sound practices. That is appropriate church counsel, but no one ever reads church convention resolutions. However, many of the **CEO**s of **COMPANIES, CLERGY,** and **CAPTIVE CONGREGANTS** all go through the same doors week after week, and in many cases year after year. Publishing, in bold type, resolutions affecting our lives, the health of creation and our children, on the front doors of our houses of worship would have the same effect as the magnet on the refrigerator door that reminds us of our important events or tasks. Compliance with existing resolutions, regulations, laws and scriptural directives about environmental stewardship should become common practice of all communities of faith.

As a part of general policy, communities of faith elect leading citizens of the community to their boards of directors. These men and women often control production in corporations and companies or run small businesses.   Together with their clergy, they could draft congregational environmental standards and conversion procedures that could save money and benefit the environment at the same time. Synagogues and churches that create environmentally ethical models of congregational life will lead in Oikonomia, the management of the household of God. Energy audits, lighting retrofitting projects, conversion to recycled paper for bulletins and leaflets, and sound environmental product purchasing procedures would model this ethic. Thus, congregants in communities of faith are a prime source of the emerging team of eco-conservatives. The marriage of business and spiritual leaders is essential to the restoration of the Garden of Eden. Management of God's Kingdom is an environmental issue that needs good business counsel and communities of faith are filled with wise persons.

The Environmental Minister from Uruguay offered a simple **CHALLENGE** to the **CLERGY** throughout the world when he said: "Unfortunately the rapid advances of science have not been accompanied by a rapid advance in thinking. The world faces a crisis of values, from which a new world order must emerge. If this does not happen, a solution to the environmental problem will not be found." Clergy and spiritual leaders are often respected as the guides to common ground and common values in their cultures. Therefore, because of the urgency of the environmental message today, sermons and worship services must contain stories or thoughts connecting to the theme of the Earth, environmental ethics, creation spirituality and practical changes. Technology is not going to solve the environmental crisis by itself. It needs a strong willed grass roots effort that will foster the conversion of the heart and that is part of the job of our communities of faith.

One guide that I have found helpful for my fellow Christian clergy is from the Lutheran Church and is called *Caring For Creation*: Vision, Hope, Justice. My rabbi friends like the Coalition on the Environment and Jewish Life's booklet called *To Till and to Tend*. In addition, the United Nations Environment Program created an inclusive multi-faith pamphlet called *Only One Earth* that offers prayers and statements from a variety of religious traditions about their environmental ethic. In addition, The National Religious Partnership for the Environment will offer guidance and help to all communities of faith in developing programs and an environmental ethic. Thankfully, many diverse religious organizations are fostering **CURRICULUM CONVERSION** so that the spiritual message is now being woven in with the teachings from the Book of Nature. Sunday schools and Hebrew schools are converting their teaching curriculums to reflect the concern of their people. "There are two books of the revelation of God," wrote Saint Thomas Aquinas, "scripture and nature." It is time that all clergy validate the marriage of these two once again.

Sustaining God's gift of the Garden will not be achieved solely via prayer, beautiful liturgy, and well-crafted sermons, but through hard work, careful reading, and challenging the systems that destroy the creatures of God. Clergy are ordained as caretakers, stewards and guarantors of God's Kingdom on Earth. There is no greater **CALLING** than **CARING** for **CREATION** and thankfully the words and deeds of persons like Reverend Carla Berkedahl, Reverend Ken Brown, Rev. Al Cohen, Dr. Don Conroy, Rev. Job Ebenezer, Fr. Matthew Fox, Fr. Jeff Golliher, Reverend Fred Kreuger, Sister Miriam T. MacGillis, Fr. Dan Martin, Pastor Peter Moore-Kochlacs, Rev. Bill Somplatsky-Jarman, Rabbi Dan Swartz, and other clergy and laypersons are beginning to penetrate the hearts of people everywhere. **CHALLENGE** your **CLERGY** to adopt an environmental ethic, and if you do not get results, give me a call and I'll give them a call....collect. It is essential to monitor the response of the clergy to the environmental crisis.

CONCLUSION

Also, if the human family watches and listens to what happens in **CHINA, CHILI, COSTA RICA,** the **CONGO, CAIRO,** the **CHESAPEAKE BAY** and **CALIFORNIA** from 1995 to the year 2000 we will most likely be able to predict the future health of the planet. These diverse **C** locations are key indicator eco systems that will continue to sound the environmental alarm bells that VP Al Gore said must begin to sound in our hearts and souls.

CHINA

China is home to more people than any other country in the world and continues to grow by 1.1 million every month. That number represents 12% of the total births of the entire planet. China is in jeopardy of not being able to feed itself. This fact has long term consequences for the stability of the world food market. Grain producing nations like ours and Canada will be stretched to the limits of our capabilities if China continues to grow at its present rate. The irony of this is that China has had the best over-all family planning programs from cradle to grave of any developing nation. How China manages its contraception deliverance system to the year 2000 should be watched carefully because the rest of the world will feel the impact.

CHINA is also rapidly becoming a **CONSUMER** nation. Conversion from bicycles to motor scooters to cars is happening overnight. This represents one of the great paradoxes of the current environmental debate.   Some population watchers believe that we destroy the environment faster if we control population because consumption goes up. They feel prosperity, not poverty destroys the environment at a more rapid pace. Thus, a huge population like China's becoming consumption driven will put enormous demands on oil and plastics, create larger pockets of pollution, and intensify reliance on chemicals and goods harmful to the environment.

CHINA'S CHICKENS should be watched carefully. China has been expanding its poultry business at an alarming rate. By the year 2000 they will require an extra flock of 1.3 billion chickens to satisfy their needs. This **CHICKEN CRAZE** alone will necessitate an additional 24 million tons of grain, today's equivalent of the entire yearly export of Canada. China is also raising 6 million tons of **CARP** a year, and fish need grain to grow. Marine feedlots are becoming a worldwide answer to the depletion of natural fish stocks; however, they require acres and acres of land and tons of food for the fish. They can also be major polluters of water systems because of the enormous amount of waste generated. China is also burning more **COAL** as it industrializes, thereby intensifying pollution and causing acid rain on a massive scale that is already being felt in their cities.

The booming economy of China and its expanding population foreshadow a cosmic collision course that will have profound effects around the world. China is an awakening dragon that may devour the Earth all by itself.

Radical solutions are mandatory for China. The U.S. can not **CHANGE CHINA**, but we can ban, with other nations, any products from entering China that contribute to the degradation of their environment. Simultaneously, we can open up China as a market for composting technologies, organic farming, sustainable forestry and agriculture, electric vehicles and alternative products that are less harmful to the environment. China alone could drive the engine of prosperity for millions while halting the deterioration of the environment, or **CHINA** could speed the **COLLAPSE** of **CREATION** to drastic proportions. China is thousands of miles away, but like the former USSR we must **CARE** about **CHINA.**

COSTA RICA

At the same time, we must pay attention to what happens in one of the smallest, least populated nations on the Earth. **COSTA RICA** is a Central American country the size of Missouri that is a key global warning nation because it sits on the cusp; i.e., Costa Rica can go either way. It can remain a model environmental nation, or it can succumb to the ravages of modern society and lose its quality of life overnight.

I spent three weeks of my sabbatical studies in Costa Rica, one week of which was spent in the northwestern corner of this country of volcanoes, rainforests, lakes and pristine beaches. I stayed up late into the evening on one occasion to observe the stars from our vantage point at the base of the extinct volcano at Rincon de la Vieja. This mountainous perch afforded me a view of hundreds of miles of the Costa Rican night. However, the darkness was broken by dozens of fires, such as I witnessed in the 1992 civil unrest in Los Angeles. However, these fires were not set out of anger, but born of presumed necessity. The clearing of forests continues in Costa Rica, as it does in many parts of the world, for the purpose of raising more cattle to produce more beef for the fast food industry. Yet, at the same time Costa Rica has designated 11% of its land mass for ecological preservation and their parks protect 75% of their total species.

Costa Rica's commitment to ecological tourism is strong, but with increased visitors comes the need for automobiles, hotels, sewage systems and people to serve the tourists. Maintaining a balance will be difficult. Watching Costa Rica closely can help us understand how eco-tourism, economic development, and environmental stewardship can go hand in hand.

Ecological tourism is big business today, but we all must learn to walk more lightly in another person's country. We all desire to see turtles, lions, or birds in the rainforests, in part because they are disappearing; but the human presence is affecting eco system survival rates worldwide. A helpful solution might be that when we acquire our passport or tourist visa we could also be given a guide for environmentally responsible tourism.

CHILI

Chili bears watching carefully for a very specific reason. The effects of the release of ozone-depleting chemicals seems to show up first in this country. The "hole in the sky" appears over the Antarctic and Chili. This has led to an increase in skin cancer, a weakening of the human immune system, and animals going blind. Chili may be far away, but it indicates an alarming trend.

Chili has been a leader in economic growth over the past two decades. However, the paradox is that Santiago, the capitol city, is now one of the worst polluted cities in the world. **CHILI** can teach the world the value of raising the standard of living of its people, but it must also teach that long term health relates to not sacrificing clean water and clean air while development occurs.

CAIRO

Halfway across the globe a challenging situation has arisen for the people of the mega-city of Cairo, Egypt. Designed to accommodate a population of 2 million, Cairo is now home to 18 million people. Watching if its infrastructure can support its population is a key indicator as to the viability of huge cities worldwide. If predictions are correct, roughly 60% of the world's population will live in megacities by the year 2000. The City of Cairo will be able to tell us a great deal about our ability to survive in **CROWDED CONGESTED CITY CORRIDORS.**

Cairo, out of necessity, has expanded dramatically, but the city is now encroaching on its symbols of antiquity. Freeways built within two miles of the pyramids may accelerate the further deterioration of those colossal monuments because of pollution and vibration. The great pyramid of **CHEOPS,** may crumble because of having to move **CARS** around the city. This is an interesting symbol of what is happening worldwide. Growth of cities impacts the surrounding eco systems. Keeping a watchful eye on Cairo's ability to handle the challenge to the year 2000 may enable us to design new ways of making mega cities sustainable.

CONGO

The Congo basin rainforest which is at the connecting corner of Cameroon and the Central African Republic is the second largest rainforest on Earth and is a forgotten place. This area is disappearing as fast, if not faster, than the Amazon Rainforest because of encroaching populations, clearing wood for fuel, and grazing cattle. Today worldwide attention is beginning to focus on the Congo;

however, poorly planned aid projects, political instability in the region, poverty, high unemployment and population pressures have placed the Congo Basin and its majestic forests in peril.

It must be noted at this point that we often feel disconnected from the problems around the world and it is hard enough just to focus on our day to day living. Many people feel frustrated and overwhelmed. Trying to focus on The Congo is a good example of this. However, as parents concerned about the long term health of our children we must support organizations working to heal damaged areas around the globe. I can stop buying any products from rainforests, except those that are harvested in a sustainable manner and that benefit the indigenous peoples. I can talk about the fact that the destruction of the rainforest there eventually will affect my life, or my children's lives here. Healing and restoring far away places ultimately helps in the healing of the whole Earth puzzle.

CHESAPEAKE BAY

Closer to home and perhaps not recognized for its environmental significance is the Chesapeake Bay. This body of water possesses a longer coastline than that of the entire United States. Its vast network of inlets, rivers, estuaries and wetlands is the spawning ground for many species, including the magnificent Atlantic Coast striped bass. Yet, like so many other places, writes William Reilly, former head of the EPA, many changes have affected the Chesapeake: "One hundred years ago, the Chesapeake Bay was teeming with oysters, plentiful enough to filter its enormous volume of water every four to five days. Now, with far fewer oysters, the same feat takes them more than a year." Because of pollution the shad and sturgeon populations are decimated, and sport fisherman can only keep striped bass over 36 inches in length because of the poor spawning record over the past ten years.

But there is good news. The Chesapeake Bay Foundation has a specific purpose and agenda. Every homeowner on the bay, every fisherman within 100 miles, and every boater, clammer, crabber and oysterman can become a Bay Watcher. In essence, the Chesapeake Bay Foundation wants to empower everyone to watch and help conserve the coastal ecology of the bay. This is an excellent model for every community in the country. Watching our own backyards and becoming community activists makes a difference. The Chesapeake represents the many thousands of water systems around the world. The successes and failures here over the next five years may signal the fate of sea creatures, sea grasses and aquatic life in many parts of the world.

CALIFORNIA

California is the last of the seven key indicator eco-systems, and brings us back to the beginning of our eco-conservative's guide to the environmental crisis. Each of us must work to conserve our own home area. We cannot focus on everything, but we can focus on California if we are Californians or we can conserve the resources of Connecticut if we live there. I have had the privilege of living in California for twenty years, and believe that the next five will set the stage for the next five hundred. Therefore, as a caring California resident my commitment to **CALIFORNIA'S** environmental health must be strong.

"This California is a hell of a mess and getting worse. We live in Ugly California," wrote William Bronson in *How to Kill a Golden State* written in 1968. Almost 30 years later, the challenge to Clean Up California bears repeating over and over again. Randy Keller, Alaskan wilderness guide, speaking at a conference in Malibu in the fall of 1994 remarked, "I put up a moose and some fish and I live for the year. I've been here in L.A. for three months and I am sick and I can't breathe."

The problem is many children from California do not really know what clean air and water smells and tastes like. California has an economic base that ranks seventh in the world, and we probably have a population consisting of every nationality represented in the United Nations. California, from Oregon to the Mexican border comprises a diversity of eco-systems that staggers the imagination. It consists of deserts, mountains, sand dunes, wetlands, cloudforests, redwood groves, and a mixture of city environments that reflect the megacities of the world. California has a combination of the best creation has to offer and the worst the human family has made. Sadly, many of our urban children have suffered from the problems but have not enjoyed the benefits. California children need us to help them enjoy their state.

One of the key indicators to watch in California is population growth. 7 of the 10 fastest growing cities in the United States are in California and our fertility rate leads the U.S. In addition, the many freeways are congested and crowded, infrastructures are crumbling, the air quality in some cities is the worst in the nation, the water delivery system is inadequate, many bays and estuaries are polluted, and yet California is still considered by millions to be the best place to live in America. What is even more remarkable is that there are thousands of people willing to convert themselves and **CONVERT CALIFORNIA** to a healthy, sustainable ecological model for the nation. City planners, government officials, environmental activists and business leaders believe in California and are creating some of the most innovative programs to **CLEAN** up **CALIFORNIA.** Cutting edge technologies, cooperative coalitions, and commitment from many diverse groups is making a difference. California represents the crowning achievements of modern society and the worst mistakes. It is not stretching the point at all to say that as California goes so goes the future of creation. If we watch California carefully to the year 2000 we will have a clear indication of whether or not the **CHALLENGE** of **COUSTEAU** and the **CHALLENGE** of the **CHILDREN** has been met. If California solves the problems raised by each of the Killer C's the rest of the world will have a reason for optimism.

CAPE COD

I close with my personal indicator of the health of the environment. I have spent a portion of every summer since 1942 in Harwichport, Massachusetts. During that time I have witnessed the disappearance of blow fish and lobster from in front of our home, the contamination of blue fish, the elimination of 3 varieties of clams, the polluting of beaches with medical waste, the dramatic decline of tautog in Cape Cod Bay, and the leaching of oil and gas into the ecological treasures of The Cape.

Visiting the exact same spot for 52 straight years has given me a context for assessing the ecology's health or damage. Because of what I know and have seen with my own eyes, I am concerned. Yet I remain hope filled because people in ever growing numbers realize the necessity of an environmentally healthy Cape Cod. I will go back to the cut at Monomoy Island to see the seals, the harbor at Provincetown to see the off loading of cod and pollack, the herring run in Brewster, the clam flats in Chatham, and the eel grass in Barnstable and Wellfleet again and again. They will inform me, teach me, warn me, and encourage me to be a person of hope. It will be easy for me to tell the health of the whole from this piece of creation's puzzle. I will stay open to the teachings of the book of nature from that special place.

CAPE COD and **CALIFORNIA** are the two most important personal Kreitler indicator C's, because this is where I live and raise my family. My voice needs to join with the protectors of these important eco-systems, the same way your voice needs to join in the protection of the eco-system where you live, work and play. Perhaps one day our collective voices will join together and form a chorus of concern for the children of all the planets' eco systems.

CHILDREN'S CHALLENGE

A commitment to the children of the world is the foundational reason for developing an environmental ethic. As adults we will witness the dawn of the year 2000 with relatively few perceptible differences. What we will see, however, should enable us to predict what our children and grandchildren will face in the future. They are the inheritors of our policies today, and the next five years will indicate to them if we have made the u-turn to a restored and sustainable creation.

So we close with one more **CHALLENGE.** We began with Cousteau's words from Rio: "You have an extra-ordinary opportunity to change the course of the world, but only if you decide to challenge the huge problems with radical solutions." We conclude with the words of a young woman from Canada which were delivered to the same delegates at the UN. I have excerpted parts of her speech and call it the **"CHILDREN'S CHALLENGE."** "Hello, I'm Severn Suzuki. I am fighting for my future. I am here to speak for all generations yet to come. I am here to speak on behalf of the starving children around the world whose cries go unheard. I'm only a child and I don't have all the solutions, but I want you to realize: neither do you! You don't know how to fix the holes in our ozone layer. You don't know how to bring the salmon back up a dead stream. You don't know how to bring back an animal now extinct. And you can't bring back the forests that once grew where there is now desert. If you don't know how to fix it, please stop breaking it! You grown-ups say you love us. I challenge you: please make your actions reflect your words. Thank you for listening."

This challenge is from the heart and mind of a 12-year-old young woman. It was delivered to people in positions of power at an historical environmental conference. A final action would be to send Severn's challenge to every leader in the world and every member of Congress asking them to act on their commitment to the children.

My personal challenge came from a 12-year-old speaking on behalf of all children to their parents. It is perhaps ironic that this is the same age as a child who went into a temple 2000 years ago and challenged the adults with his wisdom. Perhaps we will continue to choose not to listen to the ancient wisdom keepers, but we cannot afford to discount the wisdom of the 12-year-old children speaking to us today, whether in the privacy of our own homes and automobiles, or in front of world leaders at a UN Conference.

All adults are asked to listen to the voices of the children. We are also called to condemn complacency regarding the environmental crisis. The choice is ours. Each of us can be a catalyst for change in our own environment. We must encourage one another to get off the environmental sidelines and usher in the year 2000 by providing hope for the world's children. Together we can halt the collapse of creation. Each one of us, as diverse persons connected to some part of the crumbling puzzle we call Earth, can rededicate ourselves to the initial call from Genesis to keep and serve creation. Preserving creation for generations of children to come means that we must listen to their call to us and act on their behalf. Every action we take and every decision we make has an environmental consequence for the children of today and tomorrow. There is no greater calling in life than to serve creation by heeding the cries and serving the children.

The final Healing C words for the human family are important words for everyone and can function to undergird the environmental ethic so necessary to the building of the "single sacred community."

COMPASSION and **COMMITMENT** drive us to protect those places and people we love. These two traits are most often learned through the example of others. Having a passion for the gifts of creation, beginning with our own family and neighborhood, offers hope to others. Aligning ourselves with people who are committed to the health and well being of the natural world that sustains our children makes one stronger and prepared for the long journey ahead.

In addition, it is essential that we understand that **COMMUNION** and **COMMUNITY** are what make life special. Communion with creation, touching the natural places, hearing the voices of the birds, animals, and the humble indigenous people of the world prompts within the human heart the feeling of wanting to care. Caring for creation through our own communities is the avenue we have available to us. All of us live in community and that is the place we find meaning, offer hope to one another, and can live out the **CARING COMMANDMENT** so essential to the revitalization of the environment. When one starts at home it is natural that our arena for caring expands over time. This could be the essential ingredient in a new creed shared by all in the years approaching the new century. "Please make your actions reflect your words," challenges a young woman from Canada. May we listen to the wisdom of our children and change our systems on their behalf so that all in God's creation is preserved in a sustainable, healthy Garden we call Earth.

Thanks for caring.